T0280968

Synthesis Lectures on Mathematics & Statistics

Series Editor

Steven G. Krantz, Department of Mathematics, Washington University, Saint Louis, MO, USA

This series includes titles in applied mathematics and statistics for cross-disciplinary STEM professionals, educators, researchers, and students. The series focuses on new and traditional techniques to develop mathematical knowledge and skills, an understanding of core mathematical reasoning, and the ability to utilize data in specific applications.

Mircea Neagu · Alexandru Oană

Dual Jet Geometrization for Time-Dependent Hamiltonians and Applications

 Springer

Mircea Neagu ⓘ
Transylvania University of Brașov
Brașov, Romania

Alexandru Oană ⓘ
Transylvania University of Brașov
Brașov, Romania

ISSN 1938-1743   ISSN 1938-1751 (electronic)
Synthesis Lectures on Mathematics & Statistics
ISBN 978-3-031-08887-2   ISBN 978-3-031-08885-8 (eBook)
https://doi.org/10.1007/978-3-031-08885-8

© The Editor(s) (if applicable) and The Author(s), under exclusive license to Springer Nature
Switzerland AG 2022
This work is subject to copyright. All rights are solely and exclusively licensed by the Publisher, whether the whole
or part of the material is concerned, specifically the rights of translation, reprinting, reuse of illustrations, recitation,
broadcasting, reproduction on microfilms or in any other physical way, and transmission or information storage
and retrieval, electronic adaptation, computer software, or by similar or dissimilar methodology now known or
hereafter developed.
The use of general descriptive names, registered names, trademarks, service marks, etc. in this publication does
not imply, even in the absence of a specific statement, that such names are exempt from the relevant protective
laws and regulations and therefore free for general use.
The publisher, the authors, and the editors are safe to assume that the advice and information in this book are
believed to be true and accurate at the date of publication. Neither the publisher nor the authors or the editors give
a warranty, expressed or implied, with respect to the material contained herein or for any errors or omissions that
may have been made. The publisher remains neutral with regard to jurisdictional claims in published maps and
institutional affiliations.

This Springer imprint is published by the registered company Springer Nature Switzerland AG
The registered company address is: Gewerbestrasse 11, 6330 Cham, Switzerland

In memory of Professors Gheorghe Atanasiu and Radu Miron

Preface

The geometrization of Analytical Mechanics has been studied by a lot of researchers for a long time. Such a kind of geometrization was initiated by W. R. Hamilton (1805–1865), who constructed the so-called Hamiltonian mechanics as a reformulation of the Lagrangian mechanics, by replacing in the physical study the velocities with the momenta. The equations which govern the Hamiltonian mechanics are the well-known Hamilton equations which are, via the Legendre transformations, an alternative formulation of the Euler-Lagrange equations from classical mechanics.

Hamiltonian mechanics is intimately connected with the symplectic geometry. This kind of geometry relies on a $2n$-dimensional differentiable manifold M^{2n} endowed with a closed, non-degenerate 2-form ω. An important class of symplectic manifolds is given by the cotangent bundles T^*M, whose coordinates are (x^i, p_i), equipped in canonical coordinates with the symplectic form $\omega = \sum_{i=1}^{n} dx^i \wedge dp_i$. The connection with the associated Hamiltonian H is given by the Hamiltonian vector field

$$\left(\frac{\partial H}{\partial p_i}, -\frac{\partial H}{\partial x^i} \right) = \Omega dH,$$

where

$$\Omega = \omega^{-1} = \begin{pmatrix} 0 & I_n \\ -I_n & 0 \end{pmatrix}, \, dH = \begin{pmatrix} \frac{\partial H}{\partial x^i} \\ \frac{\partial H}{\partial p_i} \end{pmatrix},$$

which lead us to the Hamilton equations that govern the Hamiltonian mechanics:

$$\frac{dx^i}{dt} = \frac{\partial H}{\partial p_i}, \frac{dp_i}{dt} = -\frac{\partial H}{\partial x^i}.$$

The basics of symplectic geometry and its applications in Hamiltonian mechanics can be found in numerous monographs, e.g., Abraham and Marsden [1], Fomenko [2], or Weinstein [3]. It is important to note that, although there are many similarities with the Riemannian geometry, the symplectic manifolds have no local invariants like curvatures or torsions.

For such a reason, in contrast, Atanasiu [4, 5] and Miron et al. [6, 7] developed on the cotangent bundles the so-called *Hamilton geometry* induced by a non-degenerate Hamiltonian H, via its fundamental vertical metrical d-tensor

$$g^{ij}(x, p) = \frac{1}{2} \frac{\partial^2 H}{\partial p_i \partial p_j},$$

which is characterized by some geometrical objects as distinguished (d-) torsions and curvatures.

We further confine to the opinion expressed by P. J. Olver in his celebrated work [8], which says that 1-jet spaces and their duals are appropriate fundamental ambient mathematical spaces used to model classical and quantum field theories. In such a physical and geometrical context, suggested by the cotangent bundle framework of Gh. Atanasiu and R. Miron et al., in this monograph, we are devoted to developing the *time-dependent covariant Hamilton geometry on dual 1-jet spaces* (in the sense of d-tensors, time-dependent semisprays of momenta, nonlinear connections, N-linear connections, d-torsions, and d-curvatures), as a natural dual jet extension of the Hamilton geometry on the cotangent bundle. The geometrical study from this monograph is achieved on the *dual 1-jet vector bundle*

$$J^{1*}(\mathbb{R}, M) \equiv \mathbb{R} \times T^*M \rightarrow \mathbb{R} \times M,$$

whose local coordinates are denoted by (t, x^i, p_i^1). Here M^n is a smooth real manifold of dimension n, whose local coordinates are $(x^i)_{i=\overline{1,n}}$. The coordinates p_i^1 are called *momenta*, and the dual 1-jet space $J^{1*}(\mathbb{R}, M)$ is called the *time-dependent phase space of momenta*. The transformations of coordinates $(t, x^i, p_i^1) \leftrightarrow (\tilde{t}, \tilde{x}^i, \tilde{p}_i^1)$, induced from $\mathbb{R} \times M$ on the dual 1-jet space $J^{1*}(\mathbb{R}, M)$, have the expressions

$$\begin{cases} \tilde{t} = \tilde{t}(t) \\ \tilde{x}^i = \tilde{x}^i(x^j) \\ \tilde{p}_i^1 = \frac{\partial x^j}{\partial \tilde{x}^i} \frac{d\tilde{t}}{dt} p_j^1, \end{cases}$$

where $d\tilde{t}/dt \neq 0$ and $\det(\partial \tilde{x}^i/\partial x^j) \neq 0$. Consequently, in our dual jet geometrical approach, we use a kind of *relativistic* time t. For instance, in the Hamiltonian approach from monograph [7], the authors use the trivial bundle $\mathbb{R} \times T^*M$ over the base cotangent space T^*M, whose coordinates induced by T^*M are (t, x^i, p_i). The changes of coordinates on the trivial bundle

$$\mathbb{R} \times T^*M \rightarrow T^*M$$

are

$$\begin{cases} \tilde{t} = t \\ \tilde{x}^i = \tilde{x}^i(x^j) \\ \tilde{p}_i = \frac{\partial x^j}{\partial \tilde{x}^i} p_j, \end{cases}$$

pointing out the *absolute* character of the time variable t.

In such a context, a time-dependent Hamiltonian is a real-valued function H on $\mathbb{R} \times T^*M$, which is also called *rheonomic* or *non-autonomous* Hamiltonian. A geometrization of these Hamiltonians was sketched by Miron, Atanasiu, and their co-workers in the works [4–7]. Following their geometrical ideas, in this book, we develop a geometrization of these Hamiltonians on the dual 1-jet space $J^{1*}(\mathbb{R}, M)$. In order to emphasize the more naturalness of our dual jet approach of time-dependent Hamilton geometry, we underline that, from a geometrical point of view, the time-dependent Lagrangian theory from [7] relies on the geometrical study of the energy action integral

$$\mathbf{E}_1(c(t)) = \int_a^b L(t, x^i(t), y^i = \dot{x}^i(t)) dt$$

which has the impediment that it is dependent on the reparametrizations $t \leftrightarrow \tilde{t}$ of the same curve c. This is because $L(t, x^i, y^i)$ is a function on the vector bundle $\mathbb{R} \times TM \to M$. This inconvenience is removed in the Finsler geometry by imposing the 1-positive homogeneity condition $L(t, x^i, \lambda y^i) = \lambda L(t, x^i, y^i)$, $\forall \lambda > 0$. The second way to remove this inconvenience of dependence of reparametrizations of the energy action integral is to use the 1-jet space $J^1(\mathbb{R}, M) \equiv \mathbb{R} \times TM$ and the energy action integral (see Balan and Neagu [9])

$$\mathbf{E}_2(c(t)) = \int_a^b L(t, x^i(t), y_1^i = \dot{x}^i(t)) \sqrt{|h_{11}(t)|} dt,$$

where $L(t, x^i, y_1^i)$ is a function on the 1-jet vector bundle $J^1(\mathbb{R}, M) \to \mathbb{R} \times M$ and h_{11} is a semi-Riemannian metric on the time manifold $\mathbb{R}$. Taking into account that, via the Legendre duality of the Hamilton spaces with the Lagrange spaces, in the book [7] is shown that the theory of Hamilton spaces has the same symmetry as the Lagrange geometry, giving thus a geometrical framework for the Hamiltonian theory of Analytical Mechanics, it follows that the more natural house for the study of the time-dependent Hamilton geometry is the dual 1-jet space $J^{1*}(\mathbb{R}, M)$ which provides an energy action integral independent by temporal reparametrizations of the same curve.

The subsequent development of the time-dependent Hamilton geometry on the 1-jet space $J^{1*}(\mathbb{R}, M)$ relies on the following geometrical constructions:

(1) The writing of the time-dependent Hamiltonian $H = p_i^1 y_1^i - L$ associated with the time-dependent Lagrangian function $L(t, x^i, y_1^i)$, via the Legendre transformations $p_i^1 = \partial L / \partial y_1^i$.

(2) The production of a natural dual jet Hamiltonian nonlinear connection N (provided only by the Hamiltonian H and intimately connected with the canonical nonlinear connection produced by the Lagrangian function L, via its Euler-Lagrange equations).

(3) The construction of a natural Cartan canonical N-linear connection $C\Gamma(N)$ on the dual 1-jet space $J^{1*}(\mathbb{R}, M)$.

(4) The computations of the adapted components of the d-torsions and d-curvatures associated with the Cartan connection $C\Gamma(N)$.

(5) The development on $J^{1*}(\mathbb{R}, M)$ of some field-like (electromagnetic-like and gravitational-like) geometrical theories depending on momenta (governed by some natural momentum Maxwell-like and Einstein-like equations), starting only from the initial given Hamiltonian H.

In this way, as an application, we study in Part II of this monograph, the *dual jet time-dependent Hamiltonian of electrodynamics* (see also [7, 10])

$$H = \frac{1}{4mc}h_{11}(t)\varphi^{ij}(x)p_i^1 p_j^1 - \frac{e}{m^2c}A_{(1)}^{(i)}(x)p_i^1 + \frac{e^2}{m^3c}F(t, x) - \mathrm{P}(t, x),$$

where $A_{(1)}^{(i)}(x)$ is a d-tensor on $J^{1*}(\mathbb{R}, M)$ having the physical meaning of a potential d-tensor of an electromagnetic field, e is the charge of the test body, $\mathrm{P}(t, x)$ is a smooth function on the product manifold $\mathbb{R} \times M$, and the function $F(t, x)$ is given by $F(t, x) = h^{11}(t)\varphi_{ij}(x)A_{(1)}^{(i)}(x)A_{(1)}^{(j)}(x)$. This Hamiltonian is important because it naturally generalizes (in a time-dependent way) the Hamiltonian that governs the physical domain of the autonomous (i.e., time-independent) electrodynamics. The geometrization associated with this time-dependent Hamiltonian will consist of a canonical nonlinear connection N, a Cartan canonical N-linear connection $C\Gamma(N)$ with its adapted d-torsions and d-curvatures, together with some field-like geometrical theories depending on momenta. All these geometrical objects and theories are provided only by the initial time-dependent Hamiltonian H.

Many other applications of our geometrical theory developed in Part I of this book are exposed in Part II. We talk about some applications in Dynamical Systems, Economy, in the study of coupled harmonic oscillator or in the geometrization of the Minkowski Hamiltonian.

The authors are indebted to thank for useful and fruitful discussions upon the topics from this monograph to the following Professors: C. Udriște, V. Balan, E. M. Ovsiyuk, V. M. Red'kov, D. G. Pavlov, M. Anastasiei, I. Bucătaru, M. Păun, N. Voicu, P. Popescu, Gh. Munteanu, N. G. Krylova, H. V. Grushevskaya, M. Postolache, V. Prepeliță, A. Pitea, H. Raeisi-Dehkordi, and S. Siparov.

Brașov, Romania   Mircea Neagu
May 2022   Alexandru Oană

Contents

Part I
Time-Dependent Hamilton Geometry

The Dual 1-Jet Space $E^* = J^{1*}(\mathbb{R}, M)$

Abstract

The results of this chapter represent the basics for a subsequent geometrization on dual 1-jet spaces of the time-dependent Hamiltonians $H : J^{1*}(\mathbb{R}, M) \to \mathbb{R}$, which consist in the study of some important geometrical objects depending on momenta, such as *d-tensors*, *time-dependent semisprays* or *nonlinear connections*, together with their mathematical relations.

1.1 Dual Jet Geometrical Objects of Momenta

The present Hamilton geometrization from this chapter is similar with that developed on cotangent bundles by Miron, Atanasiu et al. [11–14], but it is characterized by the consideration of a "relativistic" time in the study. In contrast, the time-dependent Hamilton geometrization on cotangent bundles is characterized by using an absolute time. The results of this chapter follow the Neagu-Oană's paper [15].

In our geometrical study, we start with a smooth real manifold M^n of dimension n, whose local coordinates are $(x^i)_{i=\overline{1,n}}$. Let us also consider the dual 1-jet vector bundle (i.e., *the time-dependent phase space of momenta*)

$$J^{1*}(\mathbb{R}, M) \equiv \mathbb{R} \times T^*M \to \mathbb{R} \times M,$$

whose local coordinates are denoted by (t, x^i, p_i^1), where the coordinates p_i^1 have the physical meaning of *momenta*. The coordinate transformations $(t, x^i, p_i^1) \longleftrightarrow (\tilde{t}, \tilde{x}^i, \tilde{p}_i^1)$ induced from $\mathbb{R} \times M$ on the dual 1-jet space $J^{1*}(\mathbb{R}, M)$ are given by

$$\tilde{t} = \tilde{t}(t), \quad \tilde{x}^i = \tilde{x}^i(x^j), \quad \tilde{p}_i^1 = \frac{\partial x^j}{\partial \tilde{x}^i} \frac{d\tilde{t}}{dt} p_j^1, \tag{1.1}$$

© The Author(s), under exclusive license to Springer Nature Switzerland AG 2022
M. Neagu and A. Oană, *Dual Jet Geometrization for Time-Dependent Hamiltonians and Applications*, Synthesis Lectures on Mathematics & Statistics,
https://doi.org/10.1007/978-3-031-08885-8_1

where $d\tilde{t}/dt \neq 0$ and $\det(\partial \tilde{x}^i / \partial x^j) \neq 0$. Now, doing a transformation of coordinates (1.1) on $J^{1*}(\mathbb{R}, M)$, we obtain the following results.

Proposition 1.1 *The elements of the local natural basis of vector fields*

$$\left\{ \frac{\partial}{\partial t}, \frac{\partial}{\partial x^i}, \frac{\partial}{\partial p_i^1} \right\} \subset \mathcal{X}(E^*)$$

transform by the rules

$$\frac{\partial}{\partial t} = \frac{d\tilde{t}}{dt} \frac{\partial}{\partial \tilde{t}} + \frac{\partial \tilde{p}_j^1}{\partial t} \frac{\partial}{\partial \tilde{p}_j^1},$$

$$\frac{\partial}{\partial x^i} = \frac{\partial \tilde{x}^j}{\partial x^i} \frac{\partial}{\partial \tilde{x}^j} + \frac{\partial \tilde{p}_j^1}{\partial x^i} \frac{\partial}{\partial \tilde{p}_j^1}, \qquad (1.2)$$

$$\frac{\partial}{\partial p_i^1} = \frac{\partial x^i}{\partial \tilde{x}^j} \frac{d\tilde{t}}{dt} \frac{\partial}{\partial \tilde{p}_j^1}.$$

Proposition 1.2 *The elements of the local natural basis of covector fields*

$$\{ dt, dx^i, dp_i^1 \} \subset \mathcal{X}^*(E^*)$$

transform by the rules

$$dt = \frac{dt}{d\tilde{t}} d\tilde{t},$$

$$dx^i = \frac{\partial x^i}{\partial \tilde{x}^j} d\tilde{x}^j, \qquad (1.3)$$

$$dp_i^1 = \frac{\partial p_i^1}{\partial \tilde{t}} d\tilde{t} + \frac{\partial p_i^1}{\partial \tilde{x}^j} d\tilde{x}^j + \frac{\partial \tilde{x}^j}{\partial x^i} \frac{dt}{d\tilde{t}} d\tilde{p}_j^1.$$

1.2   Time-Dependent Semisprays of Momenta

As in the book [12], a central role in our dual jet geometrical study is played by *d-tensors*.

Definition 1.1 A geometrical object $T = \left(T_{1j(1)(l)...}^{1i(k)(1)...}(t, x^r, p_r^1) \right)$ on the dual 1-jet space $J^{1*}(\mathbb{R}, M)$, whose local components change according to the rules

$$T_{1j(1)(l)...}^{1i(k)(1)...} = \tilde{T}_{1q(1)(s)...}^{1p(r)(1)...} \frac{dt}{d\tilde{t}} \frac{\partial x^i}{\partial \tilde{x}^p} \left(\frac{\partial x^k}{\partial \tilde{x}^r} \frac{d\tilde{t}}{dt} \right) \frac{d\tilde{t}}{dt} \frac{\partial \tilde{x}^q}{\partial x^j} \left(\frac{\partial \tilde{x}^s}{\partial x^l} \frac{dt}{d\tilde{t}} \right) \cdots$$

with respect to a transformation of coordinates (1.1) on $J^{1*}(\mathbb{R}, M)$, is called a d-**tensor** or a **distinguished tensor field** on $J^{1*}(\mathbb{R}, M)$.

Remark 1.1 The placing between parentheses of certain indices of the local components $T^{1i(k)(1)...}_{1j(1)(l)...}$ is necessary for clearer future contractions.

Example

If $H : J^{1*}(\mathbb{R}, M) \to \mathbb{R}$ is a Hamiltonian function depending on the momenta p^1_i, then the local components

$$G^{(i)(j)}_{(1)(1)} = \frac{1}{2} \frac{\partial^2 H}{\partial p^1_i \partial p^1_j}$$

represent a d-tensor field $\mathbb{G} = \left(G^{(i)(j)}_{(1)(1)} \right)$ which is called the **vertical fundamental metrical d-tensor** produced by H.   ◀

Example

The distinguished tensor $\mathbb{C} = \left(\mathbb{C}^{(1)}_{(i)} \right)$, where $\mathbb{C}^{(1)}_{(i)} = p^1_i$, is called the **Liouville-Hamilton d-tensor field of momenta** on the dual 1-jet space $J^{1*}(\mathbb{R}, M)$.   ◀

Example

If $h_{11}(t)$ is a semi-Riemannian metric on $\mathbb{R}$, then the geometrical object $\mathbb{L} = \left(L^{(1)}_{(j)11} \right)$, where $L^{(1)}_{(j)11} = h_{11} p^1_j$ is called the **momentum Liouville-Hamilton d-tensor associated with the metric $h_{11}(t)$.**   ◀

Example

Using the preceding metric $h_{11}(t)$, the distinguished tensor $\mathbb{J} = \left(J^{(i)}_{(1)1j} \right)$, where $J^{(i)}_{(1)1j} = h_{11} \delta^i_j$, is called the d-**tensor of h-normalization** on the dual 1-jet space $J^{1*}(\mathbb{R}, M)$.   ◀

It is obvious that any d-tensor on $J^{1*}(\mathbb{R}, M)$ is a tensor field on $J^{1*}(\mathbb{R}, M)$. Conversely, the opposite is not true. As example, we construct two tensors on $J^{1*}(\mathbb{R}, M)$, which are not d-tensors on $J^{1*}(\mathbb{R}, M)$.

Definition 1.2 A global tensor $\underset{1}{G}$ on $J^{1*}(\mathbb{R}, M)$, locally expressed by

$$\underset{1}{G} = p^1_i dx^i \otimes \frac{\partial}{\partial t} - 2 \underset{1}{G}^{(1)}_{(j)i} dx^i \otimes \frac{\partial}{\partial p^1_j},$$

is called a **temporal semispray** on the dual 1-jet space $J^{1*}(\mathbb{R}, M)$.

Taking into account that the temporal semispray $\underset{1}{G}$ is a global tensor on $J^{1*}(\mathbb{R}, M)$, by a direct calculation, we obtain the following.

Proposition 1.3 (i) *Under a transformation of coordinates (1.1) the local components* $\underset{1}{G}^{(1)}_{(j)i}$ *of the global tensor* $\underset{1}{G}$ *change according to the rules*

$$2\underset{1}{\widetilde{G}}^{(1)}_{(k)r} = 2\underset{1}{G}^{(1)}_{(j)i} \frac{d\tilde{t}}{dt} \frac{\partial x^i}{\partial \tilde{x}^r} \frac{\partial x^j}{\partial \tilde{x}^k} - \frac{\partial x^i}{\partial \tilde{x}^r} \frac{\partial \tilde{p}^1_k}{\partial t} p^1_i. \tag{1.4}$$

(ii) *Conversely, to give a temporal semispray on* $J^{1*}(\mathbb{R}, M)$ *is equivalent to give a set of local functions* $\underset{1}{G} = \left(\underset{1}{G}^{(1)}_{(j)i}\right)$ *which transform by the rules (1.4).*

Example

If $H^1_{11}(t) = (h^{11}/2)(dh_{11}/dt)$ is the Christoffel symbol of a semi-Riemannian metric $h_{11}(t)$ of the temporal manifold $\mathbb{R}$, then the local components

$$\underset{1}{\overset{0}{G}}{}^{(1)}_{(j)k} = \frac{1}{2} H^1_{11} p^1_j p^1_k \tag{1.5}$$

represent a temporal semispray $\underset{1}{\overset{0}{G}}$ on $J^{1*}(\mathbb{R}, M)$, which is called the **canonical temporal semispray associated with the metric** $h_{11}(t)$.   ◀

A second example of tensor on the dual 1-jet space $J^{1*}(\mathbb{R}, M)$, which is not a distinguished tensor, is given by

Definition 1.3 A global tensor $\underset{2}{G}$ on $J^{1*}(\mathbb{R}, M)$, locally expressed by

$$\underset{2}{G} = \delta^j_i dx^i \otimes \frac{\partial}{\partial x^j} - 2\underset{2}{G}^{(1)}_{(j)i} dx^i \otimes \frac{\partial}{\partial p^1_j},$$

is called a **spatial semispray** on the dual 1-jet space $J^{1*}(\mathbb{R}, M)$.

Like in the case of a temporal semispray, we can prove without difficulties the following statement.

Proposition 1.4 *To give a spatial semispray on* $J^{1*}(\mathbb{R}, M)$ *is equivalent to give a set of local functions* $\underset{2}{G} = \left(\underset{2}{G}^{(1)}_{(j)i}\right)$ *which transform by the rules*

$$2\tilde{G}^{(1)}_{2\,(s)k} = 2G^{(1)}_{2\,(j)i}\frac{d\tilde{t}}{dt}\frac{\partial x^i}{\partial \tilde{x}^k}\frac{\partial x^j}{\partial \tilde{x}^s} - \frac{\partial x^i}{\partial \tilde{x}^k}\frac{\partial \tilde{p}^1_s}{\partial x^i}. \tag{1.6}$$

Example

If $\gamma^i_{jk}(x)$ are the Christoffel symbols of a semi-Riemannian metric $\varphi_{ij}(x)$ of the spatial manifold M, then the local components

$$\overset{0}{\underset{2}{G}}{}^{(1)}_{(j)k} = -\frac{1}{2}\gamma^i_{jk}p^1_i \tag{1.7}$$

define a spatial semispray $\overset{0}{\underset{2}{G}}$ on the dual 1-jet space $J^{1*}(\mathbb{R}, M)$, which is called the **canonical spatial semispray associated with the metric** $\varphi_{ij}(x)$. ◄

Definition 1.4 A pair $G = \left(\underset{1}{G}, \underset{2}{G}\right)$, consisting of a temporal semispray $\underset{1}{G}$ and a spatial semispray $\underset{2}{G}$, is called a **time-dependent semispray of momenta** on the dual 1-jet space $J^{1*}(\mathbb{R}, M)$.

1.3 Nonlinear Connections and Adapted Bases

In what follows, we study the important geometrical concept of *nonlinear connection* on the dual 1-jet space $J^{1*}(\mathbb{R}, M)$, which is intimately related by the concept of time-dependent semispray.

Definition 1.5 A pair of local functions $N = \left(\underset{1}{N}^{(1)}_{(k)1}, \underset{2}{N}^{(1)}_{(k)i}\right)$ on $J^{1*}(\mathbb{R}, M)$, which transform by the rules

$$\begin{aligned}
\underset{1}{\tilde{N}}^{(1)}_{(j)1} &= \underset{1}{N}^{(1)}_{(k)1}\frac{\partial x^k}{\partial \tilde{x}^j} - \frac{dt}{d\tilde{t}}\cdot\frac{\partial \tilde{p}^1_j}{\partial t}, \\
\underset{2}{\tilde{N}}^{(1)}_{(j)r} &= \underset{2}{N}^{(1)}_{(k)i}\frac{d\tilde{t}}{dt}\frac{\partial x^k}{\partial \tilde{x}^j}\frac{\partial x^i}{\partial \tilde{x}^r} - \frac{\partial x^i}{\partial \tilde{x}^r}\frac{\partial \tilde{p}^1_j}{\partial x^i},
\end{aligned} \tag{1.8}$$

is called a **nonlinear connection** on the dual 1-jet bundle $J^{1*}(\mathbb{R}, M)$. The geometrical entity $\underset{1}{N} = \left(N^{(1)}_{(j)1}\right)$ (respectively, $\underset{2}{N} = \left(N^{(1)}_{(j)i}\right)$) is called a **temporal** (respectively, **spatial) nonlinear connection** on $J^{1*}(\mathbb{R}, M)$.

Now, let us expose the connection between the time-dependent semisprays of momenta and nonlinear connections on the dual 1-jet space $J^{1*}(\mathbb{R}, M)$. For this, let us consider that $\varphi_{ij}(x)$ is a semi-Riemannian metric on the spatial manifold M. Thus, using the transformation rules (1.4), (1.6), and (1.8) of the geometrical objects taken in study, we can easily prove the following statements.

Proposition 1.5 (i) *The connection between the temporal semisprays* $G = \left(\underset{1}{G}^{(1)}_{(j)k} \right)$ *and*

the temporal components of nonlinear connections $N_{temporal} = \left(\underset{1}{N}^{(1)}_{(r)1} \right)$ *is given by the*

relations

$$\underset{1}{N}^{(1)}_{(r)1} = \varphi^{jk} \frac{\partial \underset{1}{G}^{(1)}_{(j)k}}{\partial p_i^1} \varphi_{ir}, \qquad \underset{1}{G}^{(1)}_{(i)j} = \frac{1}{2} \underset{1}{N}^{(1)}_{(i)1} p_j^1.$$

(ii) *The connection between spatial semisprays* $G = \left(\underset{2}{G}^{(1)}_{(j)i} \right)$ *and the spatial components*

of nonlinear connections $N_{spatial} = \left(\underset{2}{N}^{(1)}_{(j)i} \right)$ *is given via the relations*

$$\underset{2}{N}^{(1)}_{(j)i} = 2 \underset{2}{G}^{(1)}_{(j)i}, \qquad \underset{2}{G}^{(1)}_{(j)i} = \frac{1}{2} \underset{2}{N}^{(1)}_{(j)i}.$$

Remark 1.2 It is obvious that on the 1-jet space $J^{1*}(\mathbb{R}, M)$ a time-dependent semispray of momenta G naturally induces a nonlinear connection N_G and vice versa, a nonlinear connection N induces a time-dependent semispray G_N. The nonlinear connection N_G is called the **canonical nonlinear connection associated with the time-dependent semispray of momenta** G and **vice versa**.

Example

The canonical nonlinear connection $\overset{0}{N} = \left(\underset{1}{\overset{0}{N}}^{(1)}_{(i)1}, \underset{2}{\overset{0}{N}}^{(1)}_{(i)j} \right)$ produced by the canonical

time-dependent semispray of momenta $\overset{0}{G} = \left(\underset{1}{\overset{0}{G}}, \underset{2}{\overset{0}{G}} \right)$ has the local components

$$\underset{1}{\overset{0}{N}}^{(1)}_{(i)1} = H_{11}^1 p_i^1, \qquad \underset{2}{\overset{0}{N}}^{(1)}_{(i)j} = -\gamma_{ij}^k p_k^1. \tag{1.9}$$

This nonlinear connection is called the **canonical nonlinear connection on the 1-jet space** $J^{1*}(\mathbb{R}, M)$, **associated with the semi-Riemannian metrics** $h_{11}(t)$ **and** $\varphi_{ij}(x)$.   ◀

Taking into account the complicated transformation rules (1.2) and (1.3), we need a *horizontal distribution* on the dual 1-jet space $E^* = J^{1*}(\mathbb{R}, M)$, in order to construct some *adapted bases of vector* and *covector fields*, whose transformation rules are simpler (tensorial ones, for instance).

In this direction, let $u^* = (t, x^i, p_i^1) \in E^*$ be an arbitrary point and let us consider the differential map

$$\pi^*_{*,u^*} : T_{u^*} E^* \to T_{(t,x)} (\mathbb{R} \times M)$$

of the canonical projection

$$\pi^* : E^* = J^{1*}(\mathbb{R}, M) \to \mathbb{R} \times M, \quad \pi^* (u^*) = (t, x),$$

together with its vector subspace $W_{u^*} = Ker\,\pi^*_{*,u^*} \subset T_{u^*} E^*$. Because the differential map π^*_{*,u^*} is a surjection, we find that we have $\dim_{\mathbb{R}} W_{u^*} = n$ and, moreover, a basis in W_{u^*} is determined by $\left\{ \left. \dfrac{\partial}{\partial p_i^1} \right|_{u^*} \right\}$.

So, the map $\mathcal{W} : u^* \in J^{1*}(\mathbb{R}, M) \to W_{u^*} \subset T_{u^*} E^*$ is a differential distribution, which is called the *vertical distribution* on the dual 1-jet space $J^{1*}(\mathbb{R}, M)$.

Definition 1.6 A differential distribution

$$\mathcal{H} : u^* \in E^* = J^{1*}(\mathbb{R}, M) \to H_{u^*} \subset T_{u^*} E^*,$$

which is supplementary to the vertical distribution $\mathcal{W}$, i.e., we have

$$T_{u^*} E^* = H_{u^*} \oplus W_{u^*}, \ \forall\, u^* \in E^* = J^{1*}(\mathbb{R}, M),$$

is called a **horizontal distribution** on the dual 1-jet space $J^{1*}(\mathbb{R}, M)$.

The above definition implies that $\dim_{\mathbb{R}} H_{u^*} = n + 1$, $\forall\, u^* \in E^* = J^{1*}(\mathbb{R}, M)$. Moreover, the Lie algebra of the vector fields $X(E^*)$ can be decomposed in the direct sum $X(E^*) = S(\mathcal{H}) \oplus S(\mathcal{W})$, where $S(\mathcal{H})$ (respectively, $S(\mathcal{W})$) is the set of differentiable sections on $\mathcal{H}$ (respectively, $\mathcal{W}$).

Supposing that $\mathcal{H}$ is a fixed horizontal distribution on $J^{1*}(\mathbb{R}, M)$, we have the isomorphism

$$\pi^*_{*,u^*} \big|_{H_{u^*}} : H_{u^*} \to T_{\pi^*(u^*)} (\mathbb{R} \times M),$$

which allows us to prove the following result:

Theorem 1.1 (i) *There exist unique linear independent horizontal vector fields* $\dfrac{\delta}{\delta t}, \dfrac{\delta}{\delta x^i} \in S(\mathcal{H})$, *having the properties*

$$\pi^* {}_* \left(\frac{\delta}{\delta t} \right) = \frac{\partial}{\partial t}, \quad \pi^* {}_* \left(\frac{\delta}{\delta x^i} \right) = \frac{\partial}{\partial x^i}. \tag{1.10}$$

(ii) *The horizontal vector fields* $\dfrac{\delta}{\delta t}$ *and* $\dfrac{\delta}{\delta x^i}$ *can be uniquely written in the form*

$$\frac{\delta}{\delta t} = \frac{\partial}{\partial t} - \underset{1}{N}{}^{(1)}_{(j)1} \frac{\partial}{\partial p^1_j}, \quad \frac{\delta}{\delta x^i} = \frac{\partial}{\partial x^i} - \underset{2}{N}{}^{(1)}_{(j)i} \frac{\partial}{\partial p^1_j}. \tag{1.11}$$

(iii) *The local coefficients* $\underset{1}{N}{}^{(1)}_{(j)1}$ *and* $\underset{2}{N}{}^{(1)}_{(j)i}$ *obey the rules (1.8) of a nonlinear connection*
N *on* $J^{1*}(\mathbb{R}, M)$.

(iv) *To give on the 1-jet space* $J^{1*}(\mathbb{R}, M)$ *a horizontal distribution* $\mathcal{H}$ *is equivalent to give a nonlinear connection* $N = \left(\underset{1}{N}{}^{(1)}_{(j)1}, \underset{2}{N}{}^{(1)}_{(j)i} \right)$.

Proof Let $\dfrac{\delta}{\delta t}, \dfrac{\delta}{\delta x^i} \in \mathcal{X}(E^*)$ be vector fields on $J^{1*}(\mathbb{R}, M)$, locally expressed by

$$\frac{\delta}{\delta t} = A^1_1 \frac{\partial}{\partial t} + A^j_1 \frac{\partial}{\partial x^j} + A^{(1)}_{(j)1} \frac{\partial}{\partial p^1_j},$$

$$\frac{\delta}{\delta x^i} = X^1_i \frac{\partial}{\partial t} + X^j_i \frac{\partial}{\partial x^j} + X^{(1)}_{(j)i} \frac{\partial}{\partial p^1_j},$$

which verify relations (1.10). Then, taking into account the local expression of the map $\pi^* {}_*$, we get

$$A^1_1 = 1, \ A^j_1 = 0, \ A^{(1)}_{(j)1} = -\underset{1}{N}{}^{(1)}_{(j)1},$$

$$X^1_i = 0, \ X^j_i = \delta^j_i, \ X^{(1)}_{(j)i} = -\underset{2}{N}{}^{(1)}_{(j)i}.$$

These equalities prove the form (1.11) of the vector fields from theorem, together with their linear independence. The uniqueness of the local coefficients $\underset{1}{N}{}^{(1)}_{(j)1}$ and $\underset{2}{N}{}^{(1)}_{(j)i}$ is obvious.

Because the vector fields $\dfrac{\delta}{\delta t}$ and $\dfrac{\delta}{\delta x^i}$ are globally defined, we deduce that a change of coordinates (1.1) on $J^{1*}(\mathbb{R}, M)$ produces a transformation of the local coefficients $\underset{1}{N}{}^{(1)}_{(j)1}$ and $\underset{2}{N}{}^{(1)}_{(j)i}$ by the rules (1.8).

Finally, starting with a set of functions

$$N = \left(\underset{1}{N}{}^{(1)}_{(j)1}, \underset{2}{N}{}^{(1)}_{(j)i} \right),$$

which satisfy the rules (1.8), we can construct the horizontal distribution $\mathcal{H}$, taking

$$H_{u^*} = Span \left\{ \frac{\delta}{\delta t}\bigg|_{u^*}, \frac{\delta}{\delta x^i}\bigg|_{u^*} \right\}.$$

The decomposition $T_{u^*}E^* = H_{u^*} \oplus W_{u^*}$ is obvious now. □

Definition 1.7 The set of the linear independent vector fields

$$\left\{ \frac{\delta}{\delta t}, \frac{\delta}{\delta x^i}, \frac{\partial}{\partial p_i^1} \right\} \subset X(E^*) \tag{1.12}$$

is called the **adapted basis of vector fields produced by the nonlinear connection** $N = \left(\underset{1}{N}, \underset{2}{N} \right)$.

With respect to the coordinate transformations (1.1), the elements of the adapted basis (1.12) have their transformation laws as tensorial ones (in contrast with the transformations rules (1.2)):

$$\frac{\delta}{\delta t} = \frac{d\tilde{t}}{dt} \frac{\delta}{\delta \tilde{t}},$$

$$\frac{\delta}{\delta x^i} = \frac{\partial \tilde{x}^j}{\partial x^i} \frac{\delta}{\delta \tilde{x}^j},$$

$$\frac{\partial}{\partial p_i^1} = \frac{d\tilde{t}}{dt} \frac{\partial x^i}{\partial \tilde{x}^j} \frac{\partial}{\partial \tilde{p}_j^1}.$$

The dual basis (of covector fields) of the adapted basis (1.12) is given by

$$\left\{ dt, dx^i, \delta p_i^1 \right\} \subset X^*(E^*), \tag{1.13}$$

where

$$\delta p_i^1 = dp_i^1 + \underset{1}{N}_{(i)1}^{(1)} dt + \underset{2}{N}_{(i)j}^{(1)} dx^j.$$

Definition 1.8 The dual basis of covector fields (1.13) is called the **adapted basis of covector fields of the nonlinear connection** $N = \left(\underset{1}{N}, \underset{2}{N} \right)$.

Moreover, with respect to transformation laws (1.1), we obtain the following tensorial transformation rules:

$$dt = \frac{dt}{d\tilde{t}} d\tilde{t},$$

$$dx^i = \frac{\partial x^i}{\partial \tilde{x}^j} d\tilde{x}^j,$$

$$\delta p_i^1 = \frac{dt}{d\tilde{t}} \frac{\partial \tilde{x}^j}{\partial x^i} \delta \tilde{p}_j^1.$$

As a consequence of the preceding assertions, we find the following simple result.

Proposition 1.6 (i) *The Lie algebra of vector fields on $J^{1*}(\mathbb{R}, M)$ decomposes in the direct sum $X(E^*) = X(\mathcal{H}_{\mathbb{R}}) \oplus X(\mathcal{H}_M) \oplus X(\mathcal{W})$, where*

$$X(\mathcal{H}_{\mathbb{R}}) = Span \left\{ \frac{\delta}{\delta t} \right\}, \ X(\mathcal{H}_M) = Span \left\{ \frac{\delta}{\delta x^i} \right\}, \ X(\mathcal{W}) = Span \left\{ \frac{\partial}{\partial p_i^1} \right\}.$$

(ii) *The Lie algebra of covector fields on $J^{1*}(\mathbb{R}, M)$ decomposes in the direct sum $X^*(E^*) = X^*(\mathcal{H}_{\mathbb{R}}) \oplus X^*(\mathcal{H}_M) \oplus X^*(\mathcal{W})$, where*

$$X^*(\mathcal{H}_{\mathbb{R}}) = Span \{dt\}, \ X^*(\mathcal{H}_M) = Span \left\{ dx^i \right\}, \ X^*(\mathcal{W}) = Span \left\{ \delta p_i^1 \right\}.$$

Definition 1.9 The distributions $\mathcal{H}_{\mathbb{R}}$ and $\mathcal{H}_M$ are called the $\mathbb{R}$-**horizontal distribution** and M-**horizontal distribution** on $J^{1*}(\mathbb{R}, M)$.

In applications, the Poisson brackets of the d-vector fields (1.12) are very important. Consequently, by a direct calculus, we obtain the following.

Proposition 1.7 *The Poisson brackets of the d-vector fields of the adapted basis (1.12) are given by*

$$\left[\frac{\delta}{\delta t}, \frac{\delta}{\delta t} \right] = 0, \qquad\qquad \left[\frac{\delta}{\delta t}, \frac{\delta}{\delta x^k} \right] = R_{(i)1k}^{(1)} \frac{\partial}{\partial p_i^1},$$

$$\left[\frac{\delta}{\delta t}, \frac{\partial}{\partial p_k^1} \right] = B_{(i)1(1)}^{(1)(k)} \frac{\partial}{\partial p_i^1}, \quad \left[\frac{\delta}{\delta x^j}, \frac{\delta}{\delta x^k} \right] = R_{(i)jk}^{(1)} \frac{\partial}{\partial p_i^1}, \qquad\qquad (1.14)$$

$$\left[\frac{\delta}{\delta x^j}, \frac{\partial}{\partial p_k^1} \right] = B_{(i)j(1)}^{(1)(k)} \frac{\partial}{\partial p_i^1}, \quad \left[\frac{\partial}{\partial p_j^1}, \frac{\partial}{\partial p_k^1} \right] = 0,$$

where

$$R^{(1)}_{(i)1k} = \frac{\delta N^{(1)}_{1\,(i)1}}{\delta x^k} - \frac{\delta N^{(1)}_{2\,(i)k}}{\delta t}, \quad R^{(1)}_{(i)jk} = \frac{\delta N^{(1)}_{2\,(i)j}}{\delta x^k} - \frac{\delta N^{(1)}_{2\,(i)k}}{\delta x^j},$$

$$\tag{1.15}$$

$$B^{(1)\,(k)}_{(i)1(1)} = \frac{\partial N^{(1)}_{1\,(i)1}}{\partial p^1_k}, \qquad B^{(1)\,(k)}_{(i)j(1)} = \frac{\partial N^{(1)}_{2\,(i)j}}{\partial p^1_k},$$

and $N^{(1)}_{1\,(i)1}$ and $N^{(1)}_{2\,(i)j}$ are the coefficients of the given nonlinear connection N.

N-Linear Connections

2

Abstract

The chapter is dedicated to the study of the local adapted components of the N-linear connections on the dual 1-jet space $J^{1*}(\mathbb{R}, M)$, together with its local adapted torsion and curvature d-tensors.

2.1 Local Adapted Components

If we investigate the local adapted components of a linear connection on the 1-jet space $J^{1*}(\mathbb{R}, M)$, we remark that we have 27 components. Because of the big number of these components, we introduce the concept of N-linear connection on the dual 1-jet space $J^{1*}(\mathbb{R}, M)$, which has a reduced number of local adapted components. Its derived local adapted torsion and curvature d-tensors are also exposed. For similarities and also main differences between Miron's theory and our theory, please compare Miron et al. [16] and Oană-Neagu [17].

A linear connection on $E^* = J^{1*}(\mathbb{R}, M)$ is an application

$$D : \mathcal{X}(E^*) \times \mathcal{X}(E^*) \to \mathcal{X}(E^*), \quad (X, Y) \to D_X Y,$$

having the properties:

(1) $D_{X_1+X_2} Y = D_{X_1} Y + D_{X_2} Y,$

(2) $D_{fX} Y = f D_X Y,$

(3) $D_X (Y_1 + Y_2) = D_X Y_1 + D_X Y_2,$

(4) $D_X (fY) = X(f)Y + f D_X Y,$

© The Author(s), under exclusive license to Springer Nature Switzerland AG 2022
M. Neagu and A. Oană, *Dual Jet Geometrization for Time-Dependent Hamiltonians and Applications*, Synthesis Lectures on Mathematics & Statistics,
https://doi.org/10.1007/978-3-031-08885-8_2

where X, X_1, X_2, Y_1, Y_2, $Y \in X(E^*)$, and $f \in \mathcal{F}(E^*)$.

Obviously, the linear connection D on E^* can be uniquely determined by 27 local coefficients, which are written in the adapted basis (1.12) in the following form:

$$D_{\frac{\delta}{\delta t}} \frac{\delta}{\delta t} = A^1_{11} \frac{\delta}{\delta t} + A^i_{11} \frac{\delta}{\delta x^i} + A^{(1)}_{(i)11} \frac{\partial}{\partial p^1_i}, \tag{2.1}$$

$$D_{\frac{\delta}{\delta t}} \frac{\delta}{\delta x^j} = A^1_{j1} \frac{\delta}{\delta t} + A^i_{j1} \frac{\delta}{\delta x^i} + A^{(1)}_{(i)j1} \frac{\partial}{\partial p^1_i},$$

$$-D_{\frac{\delta}{\delta t}} \frac{\partial}{\partial p^1_j} = A^{1(j)}_{(1)1} \frac{\delta}{\delta t} + A^{i(j)}_{(1)1} \frac{\delta}{\delta x^i} + A^{(1)(j)}_{(i)(1)1} \frac{\partial}{\partial p^1_i},$$

$$D_{\frac{\delta}{\delta x^k}} \frac{\delta}{\delta t} = H^1_{1k} \frac{\delta}{\delta t} + H^i_{1k} \frac{\delta}{\delta x^i} + H^{(1)}_{(i)1k} \frac{\partial}{\partial p^1_i}, \tag{2.2}$$

$$D_{\frac{\delta}{\delta x^k}} \frac{\delta}{\delta x^j} = H^1_{jk} \frac{\delta}{\delta t} + H^i_{jk} \frac{\delta}{\delta x^i} + H^{(1)}_{(i)jk} \frac{\partial}{\partial p^1_i},$$

$$-D_{\frac{\delta}{\delta x^k}} \frac{\partial}{\partial p^1_j} = H^{1(j)}_{(1)k} \frac{\delta}{\delta t} + H^{i(j)}_{(1)k} \frac{\delta}{\delta x^i} + H^{(1)(j)}_{(i)(1)k} \frac{\partial}{\partial p^1_i},$$

$$D_{\frac{\partial}{\partial p^1_k}} \frac{\delta}{\delta t} = C^{1(k)}_{1(1)} \frac{\delta}{\delta t} + C^{i(k)}_{1(1)} \frac{\delta}{\delta x^i} + C^{(1)(k)}_{(i)1(1)} \frac{\partial}{\partial p^1_i}, \tag{2.3}$$

$$D_{\frac{\partial}{\partial p^1_k}} \frac{\delta}{\delta x^j} = C^{1(k)}_{j(1)} \frac{\delta}{\delta t} + C^{i(k)}_{j(1)} \frac{\delta}{\delta x^i} + C^{(1)(k)}_{(i)j(1)} \frac{\partial}{\partial p^1_i},$$

$$-D_{\frac{\partial}{\partial p^1_k}} \frac{\partial}{\partial p^1_j} = C^{1(j)(k)}_{(1)(1)} \frac{\delta}{\delta t} + C^{i(j)(k)}_{(1)(1)} \frac{\delta}{\delta x^i} + C^{(1)(j)(k)}_{(i)(1)(1)} \frac{\partial}{\partial p^1_i}.$$

The big number of the adapted coefficients lead us to construct linear connections whose number of coefficients is less. In this direction, let us consider a nonlinear connection N on E^*.

Definition 2.1 A linear connection D on E^* is called a N-**linear connection** if it preserves by parallelism the $\mathbb{R}$-horizontal, M-horizontal and vertical distributions $\mathcal{H}_{\mathbb{R}}$, $\mathcal{H}_M$, and $\mathcal{W}$ on E^*.

It is obvious that now a N-linear connection is uniquely described by the adapted basis of vector fields on E^* with *nine* adapted coefficients given by the following relations:

$$
D_{\frac{\delta}{\delta t}} \frac{\delta}{\delta t} = A_{11}^1 \frac{\delta}{\delta t}, \; D_{\frac{\delta}{\delta t}} \frac{\delta}{\delta x^j} = A_{j1}^i \frac{\delta}{\delta x^i}, \; D_{\frac{\delta}{\delta t}} \frac{\partial}{\partial p_j^1} = -A_{(i)(1)1}^{(1)(j)} \frac{\partial}{\partial p_i^1},
$$

$$
D_{\frac{\delta}{\delta x^k}} \frac{\delta}{\delta t} = H_{1k}^1 \frac{\delta}{\delta t}, \; D_{\frac{\delta}{\delta x^k}} \frac{\delta}{\delta x^j} = H_{jk}^i \frac{\delta}{\delta x^i}, \; D_{\frac{\delta}{\delta x^k}} \frac{\partial}{\partial p_j^1} = -H_{(i)(1)k}^{(1)(j)} \frac{\partial}{\partial p_i^1},
$$

$$
D_{\frac{\partial}{\partial p_k^1}} \frac{\delta}{\delta t} = C_{1(1)}^{1(k)} \frac{\delta}{\delta t}, \; D_{\frac{\partial}{\partial p_k^1}} \frac{\delta}{\delta x^j} = C_{j(1)}^{i(k)} \frac{\delta}{\delta x^i}, \; D_{\frac{\partial}{\partial p_k^1}} \frac{\partial}{\partial p_j^1} = -C_{(i)(1)(1)}^{(1)(j)(k)} \frac{\partial}{\partial p_i^1}.
$$

Definition 2.2 The local functions

$$
D\Gamma(N) = \left(A_{11}^1, A_{j1}^i, -A_{(i)(1)1}^{(1)(j)}, H_{1k}^1, H_{jk}^i, -H_{(i)(1)k}^{(1)(j)}, \\
C_{1(1)}^{1(k)}, C_{j(1)}^{i(k)}, -C_{(i)(1)(1)}^{(1)(j)(k)} \right) \tag{2.4}
$$

are called the **adapted coefficients of the N-linear connection** D on E^*.

Taking into account the tensorial transformation laws of the d-vector fields of the adapted basis (1.12), by direct calculations, we obtain

Theorem 2.1 (i) *With respect to the coordinate transformations (1.1) on E^*, the adapted coefficients of the N-linear connection $D\Gamma(N)$ obey the following transformation rules:*

$$
(h_{\mathbb{R}}) \begin{cases} A_{11}^1 = \tilde{A}_{11}^1 \dfrac{d\tilde{t}}{dt} + \dfrac{dt}{d\tilde{t}} \dfrac{d^2\tilde{t}}{dt^2}, \\[2mm] A_{j1}^i = \tilde{A}_{l1}^k \dfrac{\partial x^i}{\partial \tilde{x}^k} \dfrac{\partial \tilde{x}^l}{\partial x^j} \dfrac{d\tilde{t}}{dt}, \\[2mm] A_{(i)(1)1}^{(1)(j)} = \tilde{A}_{(k)(1)1}^{(1)(l)} \dfrac{\partial \tilde{x}^k}{\partial x^i} \dfrac{\partial x^j}{\partial \tilde{x}^l} \dfrac{d\tilde{t}}{dt} - \delta_i^j \dfrac{dt}{d\tilde{t}} \dfrac{d^2\tilde{t}}{dt^2}, \end{cases}
$$

$$(h_M) \begin{cases} H^1_{1k} = \tilde{H}^1_{1l} \dfrac{\partial \tilde{x}^l}{\partial x^k}, \\[2ex] H^l_{jk} = \tilde{H}^i_{rs} \dfrac{\partial \tilde{x}^r}{\partial x^j} \dfrac{\partial \tilde{x}^s}{\partial x^k} \dfrac{\partial x^l}{\partial \tilde{x}^i} + \dfrac{\partial x^l}{\partial \tilde{x}^i} \dfrac{\partial^2 \tilde{x}^i}{\partial x^j \partial x^k}, \\[2ex] H^{(1)(j)}_{(i)(1)k} = \tilde{H}^{(1)(l)}_{(r)(1)s} \dfrac{\partial \tilde{x}^r}{\partial x^i} \dfrac{\partial x^j}{\partial \tilde{x}^l} \dfrac{\partial \tilde{x}^s}{\partial x^k} - \dfrac{\partial \tilde{x}^r}{\partial x^i} \dfrac{\partial \tilde{x}^s}{\partial x^k} \dfrac{\partial^2 x^j}{\partial \tilde{x}^r \partial \tilde{x}^s}, \end{cases}$$

$$(w) \begin{cases} C^{1(k)}_{1(1)} = \tilde{C}^{1(r)}_{1(1)} \dfrac{\partial x^k}{\partial \tilde{x}^r} \dfrac{d\tilde{t}}{dt}, \\[2ex] C^{i(k)}_{j(1)} = \tilde{C}^{r(s)}_{l(1)} \dfrac{\partial x^i}{\partial \tilde{x}^r} \dfrac{\partial \tilde{x}^l}{\partial x^j} \dfrac{\partial x^k}{\partial \tilde{x}^s} \dfrac{d\tilde{t}}{dt}, \\[2ex] C^{(1)(j)(k)}_{(i)(1)(1)} = \tilde{C}^{(1)(r)(s)}_{(l)(1)(1)} \dfrac{\partial \tilde{x}^l}{\partial x^i} \dfrac{\partial x^j}{\partial \tilde{x}^r} \dfrac{\partial x^k}{\partial \tilde{x}^s} \dfrac{d\tilde{t}}{dt}. \end{cases}$$

(ii) *Conversely, to give a N-linear connection D on E^* is equivalent to give a set of nine local coefficients $D\Gamma(N)$ as in (2.4), which obey the rules described in* (i).

Example

Let us consider the canonical nonlinear connection $\overset{0}{N}$, which is given by (1.9), associated with the semi-Riemannian metrics $h_{11}(t)$ and $\varphi_{ij}(x)$. Then, the local components

$$B\Gamma\left(\overset{0}{N}\right) = \left(H^1_{11}, 0, -A^{(1)(j)}_{(i)(1)1}, 0, \gamma^i_{jk}, -H^{(1)(j)}_{(i)(1)k}, 0, 0, 0\right), \tag{2.5}$$

where

$$A^{(1)(j)}_{(i)(1)1} = -\delta^j_i H^1_{11}, \qquad H^{(1)(j)}_{(i)(1)k} = \gamma^j_{ik}, \tag{2.6}$$

define a $\overset{0}{N}$-linear connection on E^*, which is called the **canonical $\overset{0}{N}$-linear Berwald connection attached to the metrics $h_{11}(t)$ and $\varphi_{ij}(x)$.** ◄

Let us consider that D is a fixed N-linear connection on E^*, defined by the adapted coefficients (2.4). The linear connection $D\Gamma(N)$ naturally induces derivations on the set of d-tensor fields on the dual 1-jet space E^*. Starting from a d-vector field $X \in \mathcal{X}(E^*)$ and a d-tensor field T on E^*, which are locally expressed by

$$X = X^1 \frac{\delta}{\delta t} + X^i \frac{\delta}{\delta x^i} + X^{(1)}_{(i)} \frac{\partial}{\partial p^1_i},$$

$$T = T^{1i(k)(1)\dots}_{1j(1)(l)\dots} \frac{\delta}{\delta t} \otimes \frac{\delta}{\delta x^i} \otimes \frac{\partial}{\partial p^1_l} \otimes dt \otimes dx^j \otimes \delta p^1_k \otimes \dots,$$

we obtain

$$D_X T = X^1 D_{\frac{\delta}{\delta t}} T + X^s D_{\frac{\delta}{\delta x^s}} T + X^{(1)}_{(s)} D_{\frac{\partial}{\partial p_s^1}} T$$

$$= \left\{ X^1 T^{1i(k)(1)\ldots}_{1j(1)(l)\ldots/1} + X^s T^{1i(k)(1)\ldots}_{1j(1)(l)\ldots|s} + \right.$$

$$\left. + X^{(1)}_{(s)} T^{1i(k)(1)\ldots}_{1j(1)(l)\ldots} \Big|^{(s)}_{(1)} \right\} \frac{\delta}{\delta t} \otimes \frac{\delta}{\delta x^i} \otimes \frac{\partial}{\partial p_l^1} \otimes dt \otimes dx^j \otimes \delta p_k^1 \otimes \ldots,$$

where

$$(h_{\mathbb{R}}) \begin{cases} T^{1i(k)(1)\ldots}_{1j(1)(l)\ldots/1} = \dfrac{\delta T^{1i(k)(1)\ldots}_{1j(1)(l)\ldots}}{\delta t} + T^{1i(k)(1)\ldots}_{1j(1)(l)\ldots} A^1_{11} + \\[2mm] \quad + T^{1r(k)(1)\ldots}_{1j(1)(l)\ldots} A^i_{r1} + T^{1i(r)(1)\ldots}_{1j(1)(l)\ldots} A^{(1)(k)}_{(r)(1)1} + \ldots - \\[2mm] \quad - T^{1i(k)(1)\ldots}_{1j(1)(l)\ldots} A^1_{11} - T^{1i(k)(1)\ldots}_{1r(1)(l)\ldots} A^r_{j1} - \\[2mm] \quad - T^{1i(k)(1)\ldots}_{1j(1)(r)\ldots} A^{(1)(r)}_{(l)(1)1} - \ldots, \end{cases}$$

$$(h_M) \begin{cases} T^{1i(k)(1)\ldots}_{1j(1)(l)\ldots|s} = \dfrac{\delta T^{1i(k)(1)\ldots}_{1j(1)(l)\ldots}}{\delta x^s} + T^{1i(k)(1)\ldots}_{1j(1)(l)\ldots} H^1_{1s} + \\[2mm] \quad + T^{1r(k)(1)\ldots}_{1j(1)(l)\ldots} H^i_{rs} + T^{1i(r)(1)\ldots}_{1j(1)(l)\ldots} H^{(1)(k)}_{(r)(1)s} + \ldots - \\[2mm] \quad - T^{1i(k)(1)\ldots}_{1j(1)(l)\ldots} H^1_{1s} - T^{1i(k)(1)\ldots}_{1r(1)(l)\ldots} H^r_{js} - \\[2mm] \quad - T^{1i(k)(1)\ldots}_{1j(1)(r)\ldots} H^{(1)(r)}_{(l)(1)s} - \ldots, \end{cases}$$

$$(w) \begin{cases} T^{1i(k)(1)\ldots}_{1j(1)(l)\ldots} \Big|^{(s)}_{(1)} = \dfrac{\partial T^{1i(k)(1)\ldots}_{1j(1)(l)\ldots}}{\partial p_s^1} + T^{1i(k)(1)\ldots}_{1j(1)(l)\ldots} C^{1(s)}_{1(1)} + \\[2mm] \quad + T^{1r(k)(1)\ldots}_{1j(1)(l)\ldots} C^{i(s)}_{r(1)} + T^{1i(r)(1)\ldots}_{1j(1)(l)\ldots} C^{(1)(k)(s)}_{(r)(1)(1)} + \ldots - \\[2mm] \quad - T^{1i(k)(1)\ldots}_{1j(1)(l)\ldots} C^{1(s)}_{1(1)} - T^{1i(k)(1)\ldots}_{1r(1)(l)\ldots} C^{r(s)}_{j(1)} - \\[2mm] \quad - T^{1i(k)(1)\ldots}_{1j(1)(r)\ldots} C^{(1)(r)(s)}_{(l)(1)(1)} - \ldots. \end{cases}$$

Definition 2.3 The local derivative operators "$_{/1}$", "$_{|i}$", and "$|^{(i)}_{(1)}$" are called the $\mathbb{R}$-**horizontal covariant derivative**, the M-**horizontal covariant derivative**, and the **vertical covariant derivative attached to the** N-**linear connection** $D\Gamma(N)$.

Remark 2.1 The operators "$_{/1}$", "$_{|i}$", and "$|^{(i)}_{(1)}$" have the properties:

(i) They are distributive with respect to the addition of the d-tensor fields of the same type.

(ii) They commute with the operation of contraction.

(iii) They verify the Leibniz rule with respect to the tensor product.

Remark 2.2 (*i*) If $T = f(t, x^k, p_k^1)$ is a function on E^*, then the following expressions of the local covariant derivatives are true:

$$f_{/1} = \frac{\delta f}{\delta t} = \frac{\partial f}{\partial t} - \underset{1}{N}{}^{(1)}_{(i)1}\frac{\partial f}{\partial p_i^1}, \quad f_{|j} = \frac{\delta f}{\delta x^j} = \frac{\partial f}{\partial x^j} - \underset{2}{N}{}^{(1)}_{(i)j}\frac{\partial f}{\partial p_i^1},$$

$$f|^{(i)}_{(1)} = \frac{\partial f}{\partial p_i^1}.$$

(*ii*) If $T = Y$ is a *d*-vector field on E^*, locally expressed by

$$Y = Y^1\frac{\delta}{\delta t} + Y^i\frac{\delta}{\delta x^i} + Y^{(1)}_{(i)}\frac{\partial}{\partial p_i^1},$$

then the following expressions of the local covariant derivatives are true:

$$(h_{\mathbb{R}})\begin{cases} Y^1{}_{/1} = \dfrac{\delta Y^1}{\delta t} + Y^1 A^1_{11}, \\[2mm] Y^i{}_{/1} = \dfrac{\delta Y^i}{\delta t} + Y^j A^i_{j1}, \\[2mm] Y^{(1)}_{(i)/1} = \dfrac{\delta Y^{(1)}_{(i)}}{\delta t} - Y^{(1)}_{(j)} A^{(1)(j)}_{(i)(1)1}, \end{cases} \quad (h_M)\begin{cases} Y^1{}_{|k} = \dfrac{\delta Y^1}{\delta x^k} + Y^1 H^1_{1k}, \\[2mm] Y^i{}_{|k} = \dfrac{\delta Y^i}{\delta x^k} + Y^j H^i_{jk}, \\[2mm] Y^{(1)}_{(i)|k} = \dfrac{\delta Y^{(1)}_{(i)}}{\delta x^k} - Y^{(1)}_{(j)} H^{(1)(j)}_{(i)(1)k}, \end{cases}$$

$$(w)\begin{cases} Y^1\,|^{(k)}_{(1)} = \dfrac{\partial Y^1}{\partial p_k^1} + Y^1 C^{1(k)}_{1(1)}, \\[2mm] Y^i\,|^{(k)}_{(1)} = \dfrac{\partial Y^i}{\partial p_k^1} + Y^j C^{i(k)}_{j(1)}, \\[2mm] Y^{(1)}_{(i)}\,|^{(k)}_{(1)} = \dfrac{\partial Y^{(1)}_{(i)}}{\partial p_k^1} - Y^{(1)}_{(j)} C^{(1)(j)(k)}_{(i)(1)(1)}. \end{cases}$$

(*iii*) If $T = \omega$ is a *d*-covector field on E^*, locally expressed by

$$\omega = \omega_1 dt + \omega_i dx^i + \omega^{(i)}_{(1)}\delta p_i^1,$$

then the following expressions of the local covariant derivatives are true:

$$(h_{\mathbb{R}})\begin{cases} \omega_{1/1} = \dfrac{\delta \omega_1}{\delta t} - \omega_1 A^1_{11}, \\[2mm] \omega_{i/1} = \dfrac{\delta \omega_i}{\delta t} - \omega_j A^j_{i1}, \\[2mm] \omega^{(i)}_{(1)/1} = \dfrac{\delta \omega^{(i)}_{(1)}}{\delta t} + \omega^{(j)}_{(1)} A^{(1)(i)}_{(j)(1)1}, \end{cases} \quad (h_M)\begin{cases} \omega_{1|k} = \dfrac{\delta \omega_1}{\delta x^k} - \omega_1 H^1_{1k}, \\[2mm] \omega_{i|k} = \dfrac{\delta \omega_i}{\delta x^k} - \omega_j H^j_{ik}, \\[2mm] \omega^{(i)}_{(1)|k} = \dfrac{\delta \omega^{(i)}_{(1)}}{\delta x^k} + \omega^{(j)}_{(1)} H^{(1)(i)}_{(j)(1)k}, \end{cases}$$

$$
(w) \begin{cases}
\omega_1 \left.\right|_{(1)}^{(k)} = \dfrac{\partial \omega_1}{\partial p_k^1} - \omega_1 C_{1(1)}^{1(k)}, \\[2ex]
\omega_i \left.\right|_{(1)}^{(k)} = \dfrac{\partial \omega_i}{\partial p_k^1} - \omega_j C_{i(1)}^{j(k)}, \\[2ex]
\omega_{(1)}^{(i)} \left.\right|_{(1)}^{(k)} = \dfrac{\partial \omega_{(1)}^{(i)}}{\partial p_k^1} + \omega_{(1)}^{(j)} C_{(j)(1)(1)}^{(1)(i)(k)}.
\end{cases}
$$

Remark 2.3 In the particular case of the canonical Berwald $\overset{0}{N}$-linear connection given by (1.9), (2.5), and (2.6), associated with the semi-Riemannian metrics $h_{11}(t)$ and $\varphi_{ij}(x)$, the local covariant derivatives are denoted by "$_{//1}$", "$_{|i}$", and "$\left.\right|_{(1)}^{(i)}$".

Considering the canonical Liouville-Hamilton d-tensor field of momenta on E^*, which is given by

$$
\mathbb{C}^* = \mathbb{C}_{(i)}^{(1)} \frac{\partial}{\partial p_i^1} = p_i^1 \frac{\partial}{\partial p_i^1},
$$

by direct computations, we can give an application of this section

Definition 2.4 The d-tensor fields

$$
\Delta_{(i)1}^{(1)} = \mathbb{C}_{(i)/1}^{(1)}, \quad \Delta_{(i)j}^{(1)} = \mathbb{C}_{(i)|j}^{(1)}, \quad \vartheta_{(i)(1)}^{(1)(j)} = \mathbb{C}_{(i)}^{(1)}\left.\right|_{(1)}^{(j)}, \tag{2.7}
$$

are called the **momentum non-metrical deflection d-tensor fields attached to the N-linear connection** $D\Gamma(N)$.

Proposition 2.1 *The momentum deflection d-tensor fields on E^*, attached to the N-linear connection $D\Gamma(N)$, have the following expressions:*

$$
\Delta_{(i)1}^{(1)} = -\underset{1}{N}_{(i)1}^{(1)} - A_{(i)(1)1}^{(1)(k)} p_k^1, \quad \Delta_{(i)j}^{(1)} = -\underset{2}{N}_{(i)j}^{(1)} - H_{(i)(1)j}^{(1)(k)} p_k^1,
$$

$$
\vartheta_{(i)(1)}^{(1)(j)} = \delta_i^j - C_{(i)(1)(1)}^{(1)(k)(j)} p_k^1. \tag{2.8}
$$

2.2 Torsion d-Tensors

Let D be a N-linear connection on E^*. The torsion **T** of D is given by

$$
\mathbf{T}(X, Y) = D_X Y - D_Y X - [X, Y], \quad \forall X, Y \in \mathcal{X}(E^*). \tag{2.9}
$$

Let us suppose that the N-linear connection D is given in the adapted basis (1.12) by the coefficients $D\Gamma\,(N)$ from (2.4). In this context, we have

Theorem 2.2 *The local torsion d-tensors of the N-linear connection D on E^* have the following expressions:*

$$h_{\mathbb{R}}\mathbf{T}\left(\frac{\delta}{\delta t},\frac{\delta}{\delta t}\right)=T^1_{11}\frac{\delta}{\delta t},\quad h_M\mathbf{T}\left(\frac{\delta}{\delta t},\frac{\delta}{\delta t}\right)=T^k_{11}\frac{\delta}{\delta x^k},$$

$$w\mathbf{T}\left(\frac{\delta}{\delta t},\frac{\delta}{\delta t}\right)=T^{(1)}_{(r)11}\frac{\partial}{\partial p^1_r},$$

$$h_{\mathbb{R}}\mathbf{T}\left(\frac{\delta}{\delta x^j},\frac{\delta}{\delta t}\right)=T^1_{1j}\frac{\delta}{\delta t},\quad h_M\mathbf{T}\left(\frac{\delta}{\delta x^j},\frac{\delta}{\delta t}\right)=T^k_{1j}\frac{\delta}{\delta x^k},$$

$$w\mathbf{T}\left(\frac{\delta}{\delta x^j},\frac{\delta}{\delta t}\right)=T^{(1)}_{(r)1j}\frac{\partial}{\partial p^1_r},$$

$$h_{\mathbb{R}}\mathbf{T}\left(\frac{\partial}{\partial p^1_j},\frac{\delta}{\delta t}\right)=P^{1(j)}_{1(1)}\frac{\delta}{\delta t},\quad h_M\mathbf{T}\left(\frac{\partial}{\partial p^1_j},\frac{\delta}{\delta t}\right)=P^{k(j)}_{1(1)}\frac{\delta}{\delta x^k},$$

$$w\mathbf{T}\left(\frac{\partial}{\partial p^1_j},\frac{\delta}{\delta t}\right)=P^{(1)\,(j)}_{(r)1(1)}\frac{\partial}{\partial p^1_r},$$

$$h_{\mathbb{R}}\mathbf{T}\left(\frac{\delta}{\delta x^j},\frac{\delta}{\delta x^i}\right)=T^1_{ij}\frac{\delta}{\delta t},\quad h_M\,\mathbf{T}\left(\frac{\delta}{\delta x^j},\frac{\delta}{\delta x^i}\right)=T^k_{ij}\frac{\delta}{\delta x^k},$$

$$w\mathbf{T}\left(\frac{\delta}{\delta x^j},\frac{\delta}{\delta x^i}\right)=T^{(1)}_{(r)ij}\frac{\partial}{\partial p^1_r},$$

$$h_{\mathbb{R}}\mathbf{T}\left(\frac{\partial}{\partial p^1_j},\frac{\delta}{\delta x^i}\right)=P^{1(j)}_{i(1)}\frac{\delta}{\delta t},\quad h_M\mathbf{T}\left(\frac{\partial}{\partial p^1_j},\frac{\delta}{\delta x^i}\right)=P^{k(j)}_{i(1)}\frac{\delta}{\delta x^k},$$

$$w\mathbf{T}\left(\frac{\partial}{\partial p^1_j},\frac{\delta}{\delta x^i}\right)=P^{(1)\,(j)}_{(r)i(1)}\frac{\partial}{\partial p^1_r},$$

$$h_{\mathbb{R}}\mathbf{T}\left(\frac{\partial}{\partial p^1_j},\frac{\partial}{\partial p^1_i}\right)=S^{1(i)(j)}_{(1)(1)}\frac{\delta}{\delta t},\quad h_M\mathbf{T}\left(\frac{\partial}{\partial p^1_j},\frac{\partial}{\partial p^1_i}\right)=S^{k(i)(j)}_{(1)(1)}\frac{\delta}{\delta x^k},$$

$$w\mathbf{T}\left(\frac{\partial}{\partial p^1_j},\frac{\partial}{\partial p^1_i}\right)=S^{(1)(i)(j)}_{(r)(1)(1)}\frac{\partial}{\partial p^1_r},$$

where

$$\begin{cases} T^1_{11} = 0, \quad T^k_{11} = 0, \quad T^{(1)}_{(r)11} = 0, \\[4pt] T^1_{1j} = H^1_{1j}, \quad T^k_{1j} = -A^k_{j1}, \quad T^{(1)}_{(r)1j} = R^{(1)}_{(r)1j}, \\[4pt] P^{1(j)}_{1(1)} = C^{1(j)}_{1(1)}, \quad P^{k(j)}_{1(1)} = 0, \quad P^{(1)\ (j)}_{(r)1(1)} = B^{(1)\ (j)}_{(r)1(1)} + A^{(1)(j)}_{(r)(1)1}, \end{cases} \tag{2.10}$$

$$\begin{cases} T^1_{ij} = 0, \quad T^k_{ij} = H^k_{ij} - H^k_{ji}, \quad T^{(1)}_{(r)ij} = R^{(1)}_{(r)ij}, \\[4pt] T^{1(j)}_{i(1)} = 0, \quad P^{k(j)}_{i(1)} = C^{k(j)}_{i(1)}, \quad P^{(1)\ (j)}_{(r)i(1)} = B^{(1)\ (j)}_{(r)i(1)} + H^{(1)(j)}_{(r)(1)i}, \end{cases} \tag{2.11}$$

$$S^{1(i)(j)}_{(1)(1)} = 0, \quad S^{k(i)(j)}_{(1)(1)} = 0, \quad S^{(1)(i)(j)}_{(r)(1)(1)} = -\left(C^{(1)(i)(j)}_{(r)(1)(1)} - C^{(1)(j)(i)}_{(r)(1)(1)} \right), \tag{2.12}$$

and the distinguished tensors

$$R^{(1)}_{(r)1j}, \ R^{(1)}_{(r)ij}, \ B^{(1)\ (j)}_{(r)1(1)}, \ B^{(1)\ (j)}_{(r)i(1)}$$

are given by Formula (1.15).

Proof Taking into account the Poisson brackets Formulas (1.14) and (1.15), together with the local description in the adapted basis (1.12) of the N-linear connection $D\Gamma(N)$ (see (2.4)), we successively obtain

$$h_\mathbb{R} \mathbf{T} \left(\frac{\delta}{\delta t}, \frac{\delta}{\delta t} \right) = h_\mathbb{R} D_{\frac{\delta}{\delta t}} \frac{\delta}{\delta t} - h_\mathbb{R} D_{\frac{\delta}{\delta t}} \frac{\delta}{\delta t} - h_\mathbb{R} \left[\frac{\delta}{\delta t}, \frac{\delta}{\delta t} \right] = 0.$$

Consequently, the first equality from (2.10) is true. In the sequel, we have

$$h_M \mathbf{T} \left(\frac{\delta}{\delta x^j}, \frac{\delta}{\delta t} \right) = h_M D_{\frac{\delta}{\delta x^j}} \frac{\delta}{\delta t} - h_M D_{\frac{\delta}{\delta t}} \frac{\delta}{\delta x^j} - h_M \left[\frac{\delta}{\delta x^j}, \frac{\delta}{\delta t} \right] = -A^k_{j1} \frac{\delta}{\delta x^k},$$

and the fifth equality from (2.10) is correct. Then, for example, we have

$$w \mathbf{T} \left(\frac{\partial}{\partial p^1_j}, \frac{\delta}{\delta t} \right) = w D_{\frac{\partial}{\partial p^1_j}} \frac{\delta}{\delta t} - w D_{\frac{\delta}{\delta t}} \frac{\partial}{\partial p^1_j} - w \left[\frac{\partial}{\partial p^1_j}, \frac{\delta}{\delta t} \right] =$$

$$= \left(A^{(1)(j)}_{(r)(1)1} + B^{(1)\ (j)}_{(r)1(1)} \right) \frac{\partial}{\partial p^1_r},$$

and the ninth equality from (2.10) is true. In the same manner, we obtain the other equalities. □

Corollary 2.1 *The torsion* **T** *of an arbitrary N-linear connection D on E* is determined by* **ten** *effective local d-***tensors of torsion**, *arranged in Table 2.1.*

Table 2.1 d-Torsions of a N-linear connection

	$h_{\mathbb{R}}$	h_M	w
$h_{\mathbb{R}} h_{\mathbb{R}}$	0	0	0
$h_M h_{\mathbb{R}}$	T^1_{1j}	T^k_{1j}	$R^{(1)}_{(r)1j}$
$w h_{\mathbb{R}}$	$P^{1(j)}_{1(1)}$	0	$P^{(1)\,(j)}_{(r)1(1)}$
$h_M h_M$	0	T^k_{ij}	$R^{(1)}_{(r)ij}$
$w h_M$	0	$P^{k(j)}_{i(1)}$	$P^{(1)\,(j)}_{(r)i(1)}$
ww	0	0	$S^{(1)(i)(j)}_{(r)(1)(1)}$

Example

For the canonical Berwald $\overset{0}{N}$-linear connection given by (1.9), (2.5), and (2.6), associated with the semi-Riemannian metrics $h_{11}(t)$ and $\varphi_{ij}(x)$, all d-tensors of torsion vanish, except

$$R^{(1)}_{(r)ij} = -\mathfrak{R}^s_{rij}\, p^1_s,$$

where $\mathfrak{R}^s_{rij}(x)$ are the local components of the curvature tensor of the semi-Riemannian metric $\varphi_{ij}(x)$.   ◄

2.3   Curvature d-Tensors

Let D be a N-linear connection on E^*. The curvature $\mathbf{R}$ of D is given by

$$\mathbf{R}(X, Y)Z = D_X D_Y Z - D_Y D_X Z - D_{[X,Y]}Z, \quad \forall X, Y, Z \in X(E^*). \tag{2.13}$$

We will express $\mathbf{R}$ by its adapted components, taking into account the adapted local decomposition of the vector fields on E^*. In this direction, firstly we prove the following.

Theorem 2.3 *The curvature tensor field $\mathbf{R}$ of the N-linear connection D on E^* has the following properties:*

$$\begin{aligned}
&h_{\mathbb{R}}\mathbf{R}(X, Y)Z^{\mathcal{H}_M} = 0,\ h_{\mathbb{R}}\mathbf{R}(X, Y)Z^W = 0,\ h_M\mathbf{R}(X, Y)Z^{\mathcal{H}_{\mathbb{R}}} = 0,\\
&h_M\mathbf{R}(X, Y)Z^W = 0,\ w\mathbf{R}(X, Y)Z^{\mathcal{H}_{\mathbb{R}}} = 0,\ w\mathbf{R}(X, Y)Z^{\mathcal{H}_M} = 0,
\end{aligned} \tag{2.14}$$

$$\mathbf{R}(X, Y)Z = h_{\mathbb{R}}\mathbf{R}(X, Y)Z^{\mathcal{H}_{\mathbb{R}}} + h_M\mathbf{R}(X, Y)Z^{\mathcal{H}_M} + w\mathbf{R}(X, Y)Z^W. \tag{2.15}$$

Proof Because the N-linear connection D preserves by parallelism the $\mathcal{H}_{\mathbb{R}}$-horizontal, $\mathcal{H}_M$-horizontal, and vertical distributions, via Formula (2.13), the operator $\mathbf{R}(X, Y)$ carries $h_{\mathbb{R}}$-horizontal (resp., h_M-horizontal) vector fields into $h_{\mathbb{R}}$-horizontal (resp., h_M-horizontal) vector fields and the vertical vector fields into vertical vector fields. Thus, the six equations from (2.14) are true. As an easy consequence of these relations, we get (2.15). □

Taking into account the preceding geometrical result, by straightforward calculus, we obtain the following.

Theorem 2.4 *The curvature tensor* $\mathbf{R}$ *of the N-linear connection D is completely determined by* **15** *local d-tensors of curvature:*

$$\mathbf{R}\left(\frac{\delta}{\delta t}, \frac{\delta}{\delta t}\right)\frac{\delta}{\delta t} = 0, \quad \mathbf{R}\left(\frac{\delta}{\delta t}, \frac{\delta}{\delta t}\right)\frac{\delta}{\delta x^i} = 0, \quad \mathbf{R}\left(\frac{\delta}{\delta t}, \frac{\delta}{\delta t}\right)\frac{\partial}{\partial p_i^1} = 0,$$

$$\mathbf{R}\left(\frac{\delta}{\delta x^k}, \frac{\delta}{\delta t}\right)\frac{\delta}{\delta t} = R_{11k}^1\frac{\delta}{\delta t}, \quad \mathbf{R}\left(\frac{\delta}{\delta x^k}, \frac{\delta}{\delta t}\right)\frac{\delta}{\delta x^i} = R_{i1k}^l\frac{\delta}{\delta x^l},$$

$$\mathbf{R}\left(\frac{\delta}{\delta x^k}, \frac{\delta}{\delta t}\right)\frac{\partial}{\partial p_i^1} = -R_{(l)(1)1k}^{(1)(i)}\frac{\partial}{\partial p_l^1},$$

$$\mathbf{R}\left(\frac{\partial}{\partial p_k^1}, \frac{\delta}{\delta t}\right)\frac{\delta}{\delta t} = P_{11(1)}^{1\ (k)}\frac{\delta}{\delta t}, \quad \mathbf{R}\left(\frac{\partial}{\partial p_k^1}, \frac{\delta}{\delta t}\right)\frac{\delta}{\delta x^i} = P_{i1(1)}^{l\ (k)}\frac{\delta}{\delta x^l},$$

$$\mathbf{R}\left(\frac{\partial}{\partial p_k^1}, \frac{\delta}{\delta t}\right)\frac{\partial}{\partial p_i^1} = -P_{(l)(1)1(1)}^{(1)(i)\ (k)}\frac{\partial}{\partial p_l^1},$$

$$\mathbf{R}\left(\frac{\delta}{\delta x^k}, \frac{\delta}{\delta x^j}\right)\frac{\delta}{\delta t} = R_{1jk}^1\frac{\delta}{\delta t}, \quad \mathbf{R}\left(\frac{\delta}{\delta x^k}, \frac{\delta}{\delta x^j}\right)\frac{\delta}{\delta x^i} = R_{ijk}^l\frac{\delta}{\delta x^l},$$

$$\mathbf{R}\left(\frac{\delta}{\delta x^k}, \frac{\delta}{\delta x^j}\right)\frac{\partial}{\partial p_i^1} = -R_{(l)(1)jk}^{(1)(i)}\frac{\partial}{\partial p_l^1},$$

$$\mathbf{R}\left(\frac{\partial}{\partial p_k^1}, \frac{\delta}{\delta x^j}\right)\frac{\delta}{\delta t} = P_{1j(1)}^{1\ (k)}\frac{\delta}{\delta t}, \quad \mathbf{R}\left(\frac{\partial}{\partial p_k^1}, \frac{\delta}{\delta x^j}\right)\frac{\delta}{\delta x^i} = P_{ij(1)}^{l\ (k)}\frac{\delta}{\delta x^l},$$

$$\mathbf{R}\left(\frac{\partial}{\partial p_k^1}, \frac{\delta}{\delta x^j}\right)\frac{\partial}{\partial p_i^1} = -P_{(l)(1)j(1)}^{(1)(i)\ (k)}\frac{\partial}{\partial p_l^1},$$

$$\mathbf{R}\left(\frac{\partial}{\partial p_k^1}, \frac{\partial}{\partial p_j^1}\right)\frac{\delta}{\delta t} = S_{1(1)(1)}^{1(j)(k)}\frac{\delta}{\delta t}, \quad \mathbf{R}\left(\frac{\partial}{\partial p_k^1}, \frac{\partial}{\partial p_j^1}\right)\frac{\delta}{\delta x^i} = S_{i(1)(1)}^{l(j)(k)}\frac{\delta}{\delta x^l},$$

$$\mathbf{R}\left(\frac{\partial}{\partial p_k^1}, \frac{\partial}{\partial p_j^1}\right)\frac{\partial}{\partial p_i^1} = -S_{(l)(1)(1)(1)}^{(1)(i)(j)(k)}\frac{\partial}{\partial p_l^1},$$

which can be arranged in Table 2.2.

Theorem 2.5 *The 15 local curvature d-tensors from Table 2.2 are given by the following formulas:*

1. $R_{11k}^1 = \dfrac{\delta A_{11}^1}{\delta x^k} - \dfrac{\delta H_{1k}^1}{\delta t} + C_{1(1)}^{1(r)} R_{(r)1k}^{(1)},$

2. $R_{i1k}^l = \dfrac{\delta A_{i1}^l}{\delta x^k} - \dfrac{\delta H_{ik}^l}{\delta t} + A_{i1}^r H_{rk}^l - H_{ik}^r A_{r1}^l + C_{i(1)}^{l(r)} R_{(r)1k}^{(1)},$

3. $R_{(l)(1)1k}^{(1)(i)} = \dfrac{\delta A_{(l)(1)1}^{(1)(i)}}{\delta x^k} - \dfrac{\delta H_{(l)(1)k}^{(1)(i)}}{\delta t} + A_{(l)(1)1}^{(1)(r)} H_{(r)(1)k}^{(1)(i)} -$
$- H_{(l)(1)k}^{(1)(r)} A_{(r)(1)1}^{(1)(i)} + C_{(l)(1)(1)}^{(1)(i)(r)} R_{(r)1k}^{(1)},$

4. $P_{11(1)}^{1\ (k)} = \dfrac{\partial A_{11}^1}{\partial p_k^1} - C_{1(1)/1}^{1(k)} + C_{1(1)}^{1(r)} P_{(r)1(1)}^{(1)\ (k)},$

5. $P_{i1(1)}^{l\ (k)} = \dfrac{\partial A_{i1}^l}{\partial p_k^1} - C_{i(1)/1}^{l(k)} + C_{i(1)}^{l(r)} P_{(r)1(1)}^{(1)\ (k)},$

6. $P_{(l)(1)1(1)}^{(1)(i)\ (k)} = \dfrac{\partial A_{(l)(1)1}^{(1)(i)}}{\partial p_k^1} - C_{(l)(1)(1)/1}^{(1)(i)(k)} + C_{(l)(1)(1)}^{(1)(i)(r)} P_{(r)1(1)}^{(1)\ (k)},$

7. $R_{1jk}^1 = \dfrac{\delta H_{1j}^1}{\delta x^k} - \dfrac{\delta H_{1k}^1}{\delta x^j} + C_{1(1)}^{1(r)} R_{(r)jk}^{(1)},$

8. $R_{ijk}^l = \dfrac{\delta H_{ij}^l}{\delta x^k} - \dfrac{\delta H_{ik}^l}{\delta x^j} + H_{ij}^r H_{rk}^l - H_{ik}^r H_{rj}^l + C_{i(1)}^{l(r)} R_{(r)jk}^{(1)},$

9. $R_{(l)(1)jk}^{(1)(i)} = \dfrac{\delta H_{(l)(1)j}^{(1)(i)}}{\delta x^k} - \dfrac{\delta H_{(l)(1)k}^{(1)(i)}}{\delta x^j} + H_{(l)(1)j}^{(1)(r)} H_{(r)(1)k}^{(1)(i)} -$
$- H_{(l)(1)k}^{(1)(r)} H_{(r)(1)j}^{(1)(i)} + C_{(l)(1)(1)}^{(1)(i)(r)} R_{(r)jk}^{(1)},$

10. $P_{1j(1)}^{1\ (k)} = \dfrac{\partial H_{1j}^1}{\partial p_k^1} - C_{1(1)|j}^{1(k)} + C_{1(1)}^{1(r)} P_{(r)j(1)}^{(1)\ (k)},$

11. $P_{ij(1)}^{l\ (k)} = \dfrac{\partial H_{ij}^l}{\partial p_k^1} - C_{i(1)|j}^{l(k)} + C_{i(1)}^{l(r)} P_{(r)j(1)}^{(1)\ (k)},$

12. $P_{(l)(1)j(1)}^{(1)(i)\ (k)} = \dfrac{\partial H_{(l)(1)j}^{(1)(i)}}{\partial p_k^1} - C_{(l)(1)(1)|j}^{(1)(i)(k)} + C_{(l)(1)(1)}^{(1)(i)(r)} P_{(r)j(1)}^{(1)\ (k)},$

Table 2.2 d-Curvatures of a N-linear connection

	$h_\mathbb{R}$	h_M	w
$h_\mathbb{R}h_\mathbb{R}$	0	0	0
$h_M h_\mathbb{R}$	R^1_{11k}	R^l_{i1k}	$-R^{(1)(i)}_{(l)(1)1k}$
$wh_\mathbb{R}$	$P^{1\ (k)}_{11(1)}$	$P^{l\ (k)}_{i1(1)}$	$-P^{(1)(i)\ (k)}_{(l)(1)1(1)}$
$h_M h_M$	R^1_{1jk}	R^l_{ijk}	$-R^{(1)(i)}_{(l)(1)jk}$
wh_M	$P^{1\ (k)}_{1j(1)}$	$P^{l\ (k)}_{ij(1)}$	$-P^{(1)(i)\ (k)}_{(l)(1)j(1)}$
ww	$S^{1(j)(k)}_{1(1)(1)}$	$S^{l(j)(k)}_{i(1)(1)}$	$-S^{(1)(i)(j)(k)}_{(l)(1)(1)(1)}$

13. $S^{1(j)(k)}_{1(1)(1)} = \dfrac{\partial C^{1(j)}_{1(1)}}{\partial p^1_k} - \dfrac{\partial C^{1(k)}_{1(1)}}{\partial p^1_j}$,

14. $S^{l(j)(k)}_{i(1)(1)} = \dfrac{\partial C^{l(j)}_{i(1)}}{\partial p^1_k} - \dfrac{\partial C^{l(k)}_{i(1)}}{\partial p^1_j} + C^{r(j)}_{i(1)}C^{l(k)}_{r(1)} - C^{r(k)}_{i(1)}C^{l(j)}_{r(1)}$,

15. $S^{(1)(i)(j)(k)}_{(l)(1)(1)(1)} = \dfrac{\partial C^{(1)(i)(j)}_{(l)(1)(1)}}{\partial p^1_k} - \dfrac{\partial C^{(1)(i)(k)}_{(l)(1)(1)}}{\partial p^1_j} + C^{(1)(r)(j)}_{(l)(1)(1)}C^{(1)(i)(k)}_{(r)(1)(1)} -$

$\qquad\qquad\qquad - C^{(1)(r)(k)}_{(l)(1)(1)}C^{(1)(i)(j)}_{(r)(1)(1)}$.

Proof The local decomposition in the adapted basis (1.12) of the N-linear connection $D\Gamma(N)$ (see (2.4)), together with Formulas (1.14) and (1.15), leads us to, for example,

$$\mathbf{R}\left(\frac{\partial}{\partial p^1_k}, \frac{\delta}{\delta t}\right)\frac{\partial}{\partial p^1_i} = -P^{(1)(i)\ (k)}_{(l)(1)1(1)}\frac{\partial}{\partial p^1_l}$$

$$= D_{\frac{\partial}{\partial p^1_k}}D_{\frac{\delta}{\delta t}}\frac{\partial}{\partial p^1_i} - D_{\frac{\delta}{\delta t}}D_{\frac{\partial}{\partial p^1_k}}\frac{\partial}{\partial p^1_i} - D_{\left[\frac{\partial}{\partial p^1_k}, \frac{\delta}{\delta t}\right]}\frac{\partial}{\partial p^1_i}$$

$$= -D_{\frac{\partial}{\partial p^1_k}}\left(A^{(1)(i)}_{(r)(1)1}\frac{\partial}{\partial p^1_r}\right) + D_{\frac{\delta}{\delta t}}\left(C^{(1)(i)(k)}_{(r)(1)(1)}\frac{\partial}{\partial p^1_r}\right)$$

$$+ B^{(1)\ (k)}_{(r)1(1)}D_{\frac{\partial}{\partial p^1_r}}\frac{\partial}{\partial p^1_i}$$

$$= -\frac{\partial A^{(1)(i)}_{(l)(1)1}}{\partial p^1_k} \frac{\partial}{\partial p^1_l} + A^{(1)(i)}_{(r)(1)1} C^{(1)(r)(k)}_{(l)(1)(1)} \frac{\partial}{\partial p^1_l}$$

$$+ \frac{\delta C^{(1)(i)(k)}_{(l)(1)(1)}}{\delta t} \frac{\partial}{\partial p^1_l} - C^{(1)(i)(k)}_{(r)(1)(1)} A^{(1)(r)}_{(l)(1)1} \frac{\partial}{\partial p^1_l}$$

$$- B^{(1)\ (k)}_{(r)1(1)} C^{(1)(i)(r)}_{(l)(1)(1)} \frac{\partial}{\partial p^1_l}.$$

Therefore, we have

$$P^{(1)(i)\ (k)}_{(l)(1)1(1)} = \frac{\partial A^{(1)(i)}_{(l)(1)1}}{\partial p^1_k} - \underline{A^{(1)(i)}_{(r)(1)1} C^{(1)(r)(k)}_{(l)(1)(1)}} -$$

$$- \underline{\frac{\delta C^{(1)(i)(k)}_{(l)(1)(1)}}{\delta t}} + C^{(1)(i)(k)}_{(r)(1)(1)} A^{(1)(r)}_{(l)(1)1} + B^{(1)\ (k)}_{(r)1(1)} C^{(1)(i)(r)}_{(l)(1)(1)}.$$

Now, using the formula of the $\mathbb{R}$-horizontal covariant derivative, we get

$$C^{(1)(i)(k)}_{(l)(1)(1)/1} = \underline{\frac{\delta C^{(1)(i)(k)}_{(l)(1)(1)}}{\delta t}} - C^{(1)(i)(k)}_{(r)(1)(1)} A^{(1)(r)}_{(l)(1)1} + \underline{C^{(1)(r)(k)}_{(l)(1)(1)} A^{(1)(i)}_{(r)(1)1}} +$$

$$+ C^{(1)(i)(r)}_{(l)(1)(1)} A^{(1)(k)}_{(r)(1)1},$$

and, consequently, interchanging the underlined terms, it follows that

$$P^{(1)(i)\ (k)}_{(l)(1)1(1)} = \frac{\partial A^{(1)(i)}_{(l)(1)1}}{\partial p^1_k} - C^{(1)(i)(k)}_{(l)(1)(1)/1} + C^{(1)(i)(r)}_{(l)(1)(1)} P^{(1)\ (k)}_{(r)1(1)},$$

where we also used the last formula from (2.10). Obviously, this is the 6-*th* relation of the above set of identities.

The other equalities are given in the same manner.   □

Example

For the canonical Berwald $\overset{0}{N}$-linear connection given by (1.9), (2.5), and (2.6), associated with the semi-Riemannian metrics $h_{11}(t)$ and $\varphi_{ij}(x)$, all curvature *d*-tensors vanish, except

$$R^l_{ijk} = R^{(1)(l)}_{(i)(1)jk} = \mathfrak{R}^l_{ijk},$$

where $\mathfrak{R}^l_{ijk}(x)$ are the local curvature tensors of the semi-Riemannian metric $\varphi_{ij}(x)$. ◄

h-Normal N-Linear Connections

3

Abstract

In this chapter, we study the local adapted components of the h-normal N-linear connections in the dual jet time-dependent Hamilton geometry. Their corresponding local d-torsions and d-curvatures are computed. As a consequence, we describe on the dual 1-jet space $J^{1*}(\mathbb{R}, M)$ the Ricci and non-metrical deflection d-tensor identities of a h-normal N-linear connection of Cartan type. Moreover, the local Bianchi identities for a h-normal N-linear connection of Cartan type are also described.

3.1   Local Adapted Components

A general linear connection on the dual 1-jet space $J^{1*}(\mathbb{R}, M)$ is characterized by 27 adapted local components. To work with these components is not impossible, but is very complicated. For such a reason, in Chap. 2 were introduced and studied the N-linear connections. Because the number of components which characterizes a N-linear connection on $E^* = J^{1*}(\mathbb{R}, M)$ is still big one (nine local components), then we are constrained to study only a particular class of N-linear connections on E^*, which must be characterized by a reduced number of components (only four local components in this case of that so-called h-normal N-linear connections). A detailed study of such kind of connections can be found in the papers [18–20], developed by Balan, Neagu, and Oană.

In this direction, let us consider a semi-Riemannian metric $h_{11}(t)$ on the time manifold $\mathbb{R}$, together with its Christoffel symbol

$$H_{11}^1 = \frac{h^{11}}{2} \frac{dh_{11}}{dt}.$$

Let $\mathbb{J}$ be the h-normalization d-tensor field on E^*, locally expressed by

© The Author(s), under exclusive license to Springer Nature Switzerland AG 2022

M. Neagu and A. Oană, *Dual Jet Geometrization for Time-Dependent Hamiltonians and Applications*, Synthesis Lectures on Mathematics & Statistics,

https://doi.org/10.1007/978-3-031-08885-8_3

$$\mathbb{J} = J^{(i)}_{(1)1j}\delta p^1_i \otimes dt \otimes dx^j,$$

where $J^{(i)}_{(1)1j} = h_{11}\delta^i_j$. In this context, we introduce the following geometrical concept.

Definition 3.1 A N-linear connection $D\Gamma(N)$ on E^*, locally given by

$$D\Gamma(N) = \left(A^1_{11}, A^i_{j1}, -A^{(1)(j)}_{(i)(1)1}, H^1_{1k}, H^i_{jk}, -H^{(1)(j)}_{(i)(1)k},\right.$$
$$\left. C^{1(k)}_{1(1)}, C^{i(k)}_{j(1)}, -C^{(1)(j)(k)}_{(i)(1)(1)}\right),$$

whose local components verify the relations

$$A^1_{11} = H^1_{11}, \ H^1_{1i} = 0, \ C^{1(i)}_{1(1)} = 0, \ D\mathbb{J} = 0,$$

is called a h-**normal** N-**linear connection** on the dual 1-jet space E^*.

Theorem 3.1 *The local adapted components of a h-normal N-linear connection $D\Gamma(N)$ verify the following identities:*

$$A^1_{11} = H^1_{11}, \quad H^1_{1i} = 0, \quad C^{1(i)}_{1(1)} = 0,$$
$$A^{(1)(j)}_{(i)(1)1} = A^j_{i1} - \delta^j_i H^1_{11}, \quad H^{(1)(j)}_{(i)(1)k} = H^j_{ik}, \tag{3.1}$$
$$C^{(1)(j)(k)}_{(i)(1)(1)} = C^{j(k)}_{i(1)}.$$

Proof It is obvious that the first three relations immediately come from the definition of a h-normal N-linear connection. To prove the other three relations, we emphasize that, taking into account the definition of the local $\mathbb{R}$-horizontal ("$/1$"), M-horizontal ("$|_s$") and vertical ("$|^{(s)}_{(1)}$") covariant derivatives produced by $D\Gamma(N)$, the condition $D\mathbb{J} = 0$ is equivalent to

$$J^{(i)}_{(1)1j/1} = 0, \ J^{(i)}_{(1)1j|s} = 0, \ J^{(i)}_{(1)1j}|^{(s)}_{(1)} = 0.$$

Consequently, the condition $D\mathbb{J} = 0$ provides the local identities

$$h_{11}A^{(1)(i)}_{(j)(1)1} = h_{11}A^i_{j1} - \delta^i_j\left(\frac{dh_{11}}{dt} - h_{11}H^1_{11}\right) = h_{11}A^i_{j1} - \delta^i_j\frac{1}{2}\frac{dh_{11}}{dt},$$
$$h_{11}H^{(1)(i)}_{(j)(1)k} = h_{11}H^i_{jk}, \quad h_{11}C^{(1)(i)(k)}_{(j)(1)(1)} = h_{11}C^{i(k)}_{j(1)}.$$

Multiplying now the above relations by h^{11}, we obtain the last required identities from (3.1). □

Remark 3.1 The above theorem shows that a h-normal N-linear connection on E^* is a N-linear connection determined only by **four** effective components (instead of nine in the general case):

Table 3.1 d-Torsions of a h-normal N-linear connection

	$h_\mathbb{R}$	h_M	v
$h_\mathbb{R}h_\mathbb{R}$	0	0	0
$h_M h_\mathbb{R}$	0	T^r_{1j}	$R^{(1)}_{(r)1j}$
$vh_\mathbb{R}$	0	0	$P^{(1)\,(j)}_{(r)1(1)}$
$h_M h_M$	0	T^r_{ij}	$R^{(1)}_{(r)ij}$
vh_M	0	$P^{r(j)}_{i(1)}$	$P^{(1)\,(j)}_{(r)i(1)}$
vv	0	0	$S^{(1)(i)(j)}_{(r)(1)(1)}$

$$D\Gamma(N) = \left(H^1_{11},\ A^i_{j1},\ H^i_{jk},\ C^{i(k)}_{j(1)}\right).$$

The other five components either vanish or are provided by Relations (3.1).

Example

The Berwald $\overset{0}{N}$-linear connection associated with the pair of metrics $\left(h_{11}(t), \varphi_{ij}(x)\right)$ is a h-normal $\overset{0}{N}$-linear connection on E^*, whose four effective components are

$$B\Gamma\left(\overset{0}{N}\right) = \left(H^1_{11}(t),\ 0,\ \gamma^i_{jk}(x),\ 0\right). \quad\blacktriangleleft$$

3.2 Torsion and Curvature d-Tensors

The study of the adapted components of the torsion and curvature tensors of an arbitrary N-linear connection $D\Gamma(N)$ on E^* was done in the preceding sections. In that context, it turned out that the torsion tensor **T** has *ten* effective local adapted d-tensors, while the curvature tensor **R** has *15* local adapted d-tensors. In what follows, we study the adapted components of the torsion and curvature tensors for a h-normal N-linear connection $D\Gamma(N)$.

Theorem 3.2 *The torsion tensor* **T** *of a h-normal N-linear connection $D\Gamma(N)$ has* **eight** *effective local adapted d-tensors—see Table 3.1—(instead of ten in the general case), where*

$$T^r_{1j} = -A^r_{j1},\ T^r_{ij} = H^r_{ij} - H^r_{ji},\ P^{r(j)}_{i(1)} = C^{r(j)}_{i(1)},\ S^{(1)(i)(j)}_{(r)(1)(1)} = -\left(C^{i(j)}_{r(1)} - C^{j(i)}_{r(1)}\right)$$

$$P_{(r)1(1)}^{(1)\,(j)} = \frac{\partial N_{(r)1}^{(1)}}{\partial p_j^1} + A_{r1}^j - \delta_r^j H_{11}^1, \quad P_{(r)i(1)}^{(1)\,(j)} = \frac{\partial N_{(r)i}^{(1)}}{\partial p_j^1} + H_{ri}^j,$$

$$R_{(r)1j}^{(1)} = \frac{\delta N_{(r)1}^{(1)}}{\delta x^j} - \frac{\delta N_{(r)j}^{(1)}}{\delta t}, \quad R_{(r)ij}^{(1)} = \frac{\delta N_{(r)i}^{(1)}}{\delta x^j} - \frac{\delta N_{(r)j}^{(1)}}{\delta x^i}.$$

Proof By particularizing the general local expressions of the torsion tensor of a N-linear connection from Chap. 2 (having ten d-components) for a h-normal N-linear connection on E^*, we deduce that we have $T_{1j}^1 = 0$ and $P_{1(1)}^{1(k)} = 0$, while the other eight are given by the formulas from above theorem. $\qquad\square$

Remark 3.2 All torsion d-tensors of the Berwald h-normal $\overset{0}{N}$-linear connection $B\overset{0}{\Gamma}(N)$ (associated with the metrics $h_{11}(t)$ and $\varphi_{ij}(x)$) are zero, except

$$R_{(r)ij}^{(1)} = -\mathfrak{R}_{rij}^s p_s^1,$$

where $\mathfrak{R}_{rij}^s(x)$ is the local curvature of the semi-Riemannian metric $\varphi_{ij}(x)$.

Theorem 3.3 *The curvature tensor* $\mathbf{R}$ *of a h-normal N-linear connection* $D\Gamma(N)$ *has five effective adapted local d-tensors—see Table 3.2—(instead of 15 in the general case), where*

$$R_{i1k}^l = \frac{\delta A_{i1}^l}{\delta x^k} - \frac{\delta H_{ik}^l}{\delta t} + A_{i1}^r H_{rk}^l - H_{ik}^r A_{r1}^l + C_{i(1)}^{l(r)} R_{(r)1k}^{(1)},$$

$$P_{i1(1)}^{l\,(k)} = \frac{\partial A_{i1}^l}{\partial p_k^1} - C_{i(1)/1}^{l(k)} + C_{i(1)}^{l(r)} P_{(r)1(1)}^{(1)\,(k)},$$

$$R_{ijk}^l = \frac{\delta H_{ij}^l}{\delta x^k} - \frac{\delta H_{ik}^l}{\delta x^j} + H_{ij}^r H_{rk}^l - H_{ik}^r H_{rj}^l + C_{i(1)}^{l(r)} R_{(r)jk}^{(1)},$$

$$P_{ij(1)}^{l\,(k)} = \frac{\partial H_{ij}^l}{\partial p_k^1} - C_{i(1)|j}^{l(k)} + C_{i(1)}^{l(r)} P_{(r)j(1)}^{(1)\,(k)},$$

$$S_{i(1)(1)}^{l(j)(k)} = \frac{\partial C_{i(1)}^{l(j)}}{\partial p_k^1} - \frac{\partial C_{i(1)}^{l(k)}}{\partial p_j^1} + C_{i(1)}^{r(j)} C_{r(1)}^{l(k)} - C_{i(1)}^{r(k)} C_{r(1)}^{l(j)}.$$

Proof The general formulas that express the local curvature d-tensors of an arbitrary N-linear connection from Chap. 2, applied to the particular case of a h-normal N-linear connection $D\Gamma(N)$, imply the above formulas and the relations from Table 3.2. $\qquad\square$

Table 3.2 d-curvatures of a h-normal N-linear connection

	$h_{\mathbb{R}}$	h_M	v
$h_{\mathbb{R}}h_{\mathbb{R}}$	0	0	0
$h_M h_{\mathbb{R}}$	0	R^l_{i1k}	$-R^{(1)(i)}_{(l)(1)1k} = -R^i_{l1k}$
$wh_{\mathbb{R}}$	0	$P^{l\ (k)}_{i1(1)}$	$-P^{(1)(i)\ (k)}_{(l)(1)1(1)} = -P^{i\ (k)}_{l1(1)}$.
$h_M h_M$	0	R^l_{ijk}	$-R^{(1)(i)}_{(l)(1)jk} = -R^i_{ljk}$
wh_M	0	$P^{l\ (k)}_{ij(1)}$	$-P^{(1)(i)\ (k)}_{(l)(1)j(1)} = -P^{i\ (k)}_{lj(1)}$
ww	0	$S^{l(j)(k)}_{i(1)(1)}$	$-S^{(1)(i)(j)(k)}_{(l)(1)(1)(1)} = -S^{i(j)(k)}_{l(1)(1)}$

Remark 3.3 As we have already seen in Chap. 2, for the Berwald h-normal $\overset{0}{N}$-linear connection $B\Gamma(\overset{0}{N})$ (associated with the metrics $h_{11}(t)$ and $\varphi_{ij}(x)$), all curvature d-tensors are zero, except

$$R^{(1)(l)}_{(i)(1)jk} = R^l_{ijk} = \mathfrak{R}^l_{ijk},$$

where $\mathfrak{R}^s_{rij}(x)$ are the local curvature tensors of the semi-Riemannian metric $\varphi_{ij}(x)$.

3.3 Ricci and Deflection d-Tensor Identities

Let $h_{11}(t)$ be a semi-Riemannian metric on $\mathbb{R}$, together with its Christoffel symbol $H^1_{11} = \frac{h^{11}}{2}\frac{dh_{11}}{dt}$. Then, on the dual 1-jet space $E^* = J^{1*}(\mathbb{R}, M)$, a h-normal N-linear connection

$$CD\Gamma(N) = \left(H^1_{11},\ A^i_{j1},\ H^i_{jk},\ C^{i(k)}_{j(1)}\right), \tag{3.2}$$

whose local components verify the relations $H^i_{jk} = H^i_{kj}$ and $C^{i(k)}_{j(1)} = C^{k(i)}_{j(1)}$, is called a h-normal N-linear connection of Cartan type on E^*. Because the d-torsions $T^i_{jk} = H^i_{jk} - H^i_{kj}$ and $S^{(1)(i)(j)}_{(r)(1)(1)} = -\left(C^{i(j)}_{r(1)} - C^{j(i)}_{r(1)}\right)$ are zero, it follows that the torsion tensor **T** of the connection (3.2) is characterized only by *six* effective adapted local d-tensors. At the same time, its curvature tensor **R** is characterized by *five* effective adapted local d-tensors. These local torsion and curvature d-tensors are described in Tables 3.1 and 3.2.

Theorem 3.4 *The next local **Ricci identities** of the connection (3.2) are true:*

(1)—*the $h_{\mathbb{R}}$-Ricci identities:*

$$X^1_{/1|k} - X^1_{|k/1} = -X^1_{|r}T^r_{1k} - X^1|^{(r)}_{(1)}R^{(1)}_{(r)1k},$$

$$X^1_{|j|k} - X^1_{|k|j} = -X^1|^{(r)}_{(1)} R^{(1)}_{(r)jk},$$

$$X^1_{/1}|^{(k)}_{(1)} - X^1|^{(k)}_{(1)/1} = -X^1|^{(r)}_{(1)} P^{(1)}{}^{(k)}_{(r)1(1)},$$

$$X^1_{|j}|^{(k)}_{(1)} - X^1|^{(k)}_{(1)|j} = -X^1_{|r} C^{r(k)}_{j(1)} - X^1|^{(r)}_{(1)} P^{(1)}{}^{(k)}_{(r)j(1)},$$

$$X^1|^{(j)}_{(1)}|^{(k)}_{(1)} - X^1|^{(k)}_{(1)}|^{(j)}_{(1)} = 0;$$

(2)—the h_M-Ricci identities:

$$X^i_{/1|k} - X^i_{|k/1} = X^r R^i_{r1k} - X^i_{|r} T^r_{1k} - X^i|^{(r)}_{(1)} R^{(1)}_{(r)1k},$$

$$X^i_{|j|k} - X^i_{|k|j} = X^r R^i_{rjk} - X^i|^{(r)}_{(1)} R^{(1)}_{(r)jk},$$

$$X^i_{/1}|^{(k)}_{(1)} - X^i|^{(k)}_{(1)/1} = X^r P^i{}^{(k)}_{r1(1)} - X^i|^{(r)}_{(1)} P^{(1)}{}^{(k)}_{(r)1(1)},$$

$$X^i_{|j}|^{(k)}_{(1)} - X^i|^{(k)}_{(1)|j} = X^r P^i{}^{(k)}_{rj(1)} - X^i_{|r} C^{r(k)}_{j(1)} - X^i|^{(r)}_{(1)} P^{(1)}{}^{(k)}_{(r)j(1)},$$

$$X^i|^{(j)}_{(1)}|^{(k)}_{(1)} - X^i|^{(k)}_{(1)}|^{(j)}_{(1)} = X^r S^{i(j)(k)}_{r(1)(1)};$$

(3)—the v-Ricci identities:

$$X^{(1)}_{(i)/1|k} - X^{(1)}_{(i)|k/1} = X^{(1)}_{(r)} R^r_{i1k} - X^{(1)}_{(i)|r} T^r_{1k} - X^{(1)}_{(i)}|^{(r)}_{(1)} R^{(1)}_{(r)1k},$$

$$X^{(1)}_{(i)|j|k} - X^{(1)}_{(i)|k|j} = X^{(1)}_{(r)} R^r_{ijk} - X^{(1)}_{(i)}|^{(r)}_{(1)} R^{(1)}_{(r)jk},$$

$$X^{(1)}_{(i)/1}|^{(k)}_{(1)} - X^{(1)}_{(i)}|^{(k)}_{(1)/1} = X^{(1)}_{(r)} P^r{}^{(k)}_{i1(1)} - X^{(1)}_{(i)}|^{(r)}_{(1)} P^{(1)}{}^{(k)}_{(r)1(1)},$$

$$X^{(1)}_{(i)|j}|^{(k)}_{(1)} - X^{(1)}_{(i)}|^{(k)}_{(1)|j} = X^{(1)}_{(r)} P^r{}^{(k)}_{ij(1)} - X^{(1)}_{(i)|r} C^{r(k)}_{j(1)} - X^{(1)}_{(i)}|^{(r)}_{(1)} P^{(1)}{}^{(k)}_{(r)j(1)},$$

$$X^{(1)}_{(i)}|^{(j)}_{(1)}|^{(k)}_{(1)} - X^{(1)}_{(i)}|^{(k)}_{(1)}|^{(j)}_{(1)} = X^{(1)}_{(r)} S^{r(j)(k)}_{i(1)(1)},$$

where

$$X = X^1 \frac{\delta}{\delta t} + X^i \frac{\delta}{\delta x^i} + X^{(1)}_{(i)} \frac{\partial}{\partial p^1_i}$$

is an arbitrary d-vector field on the dual 1-jet space E^*.

Proof Let (Y_A) and (ω^A), where $A \in \left\{1, i, \overset{(1)}{(i)}\right\}$, be the dual bases adapted to the nonlinear connection N on E^*, and let $X = X^F Y_F$ be a d-vector field on E^*. In this context, we use the following true equalities (applied for the connection $CD\Gamma(N)$ given by (3.2))

1. $[Y_B, Y_C] = R^F_{BC} Y_F$;

2. $D_{Y_C} Y_B = \Gamma^F_{BC} Y_F$;

3. $D_{Y_C} \omega^B = -\Gamma^B_{FC} \omega^F$;

4. $\mathbf{T}(Y_C, Y_B) = T^F_{BC} Y_F = \{\Gamma^F_{BC} - \Gamma^F_{CB} - R^F_{CB}\} Y_F$;

5. $R(Y_C, Y_B)Y_A = R^F_{ABC}Y_F$;

6. $[R(Y_C, Y_B)X] \otimes \omega^B \otimes \omega^C = \{D_{Y_C}D_{Y_B}X - D_{Y_B}D_{Y_C}X - D_{[Y_C, Y_B]}X\} \otimes \omega^B \otimes \omega^C$.

It follows that, by a direct calculation, we get the equalities

$$X^A_{:B:C} - X^A_{:C:B} = X^F R^A_{FBC} - X^A_{:F} T^F_{BC}, \tag{3.3}$$

where "$:_G$" represents one from the local covariant derivatives "$_{/1}$", "$_{|j}$", or "$|^{(j)}_{(1)}$" produced by the h-normal N-linear connection of Cartan type (3.2).

Taking into account in (3.3) that the indices $A, B, C, \ldots$ belong to the set $\left\{1, i, \overset{(1)}{(i)}\right\}$, and using the particular features of the h-normal N-linear connection of Cartan type (3.2), by complicated computations, we find what we were looking for. $\qquad\square$

Now, let us consider the canonical Liouville-Hamilton d-tensor field of momenta on E^*, which is given by $\mathbb{C}^* = p^1_i \left(\partial/\partial p^1_i\right)$. In this context, for the h-normal N-linear connection of Cartan type (3.2), we construct the *non-metrical deflection d-tensors*

$$\overset{(1)}{\Delta}_{(i)1} = p^1_{i/1}, \quad \overset{(1)}{\Delta}_{(i)j} = p^1_{i|j}, \quad \overset{(1)(j)}{\vartheta}_{(i)(1)} = p^1_i |^{(j)}_{(1)},$$

where "$_{/1}$", "$_{|j}$", and "$|^{(j)}_{(1)}$" are the local covariant derivatives produced by the connection (3.2).

By local computations, we deduce that the non-metrical deflection d-tensors of the h-normal N-linear connection of Cartan type (3.2) are given by

$$\overset{(1)}{\Delta}_{(i)1} = -\underset{1}{\overset{(1)}{N}}_{(i)1} - A^r_{i1}p^1_r + H^1_{11}p^1_i,$$

$$\overset{(1)}{\Delta}_{(i)j} = -\underset{2}{\overset{(1)}{N}}_{(i)j} - H^r_{ij}p^1_r,$$

$$\overset{(1)(j)}{\vartheta}_{(i)(1)} = \delta^j_i - C^{r(j)}_{i(1)}p^1_r,$$

where

$$N = \left(\underset{1}{\overset{(1)}{N}}_{(i)1}, \underset{2}{\overset{(1)}{N}}_{(i)j}\right)$$

is the given nonlinear connection on E^*.

Applying now the preceding (v)-set of Ricci identities to the components of the canonical Liouville-Hamilton d-vector field of momenta, we find

Corollary 3.1 *The following **non-metrical deflection d-tensor identities**, associated with the h-normal N-linear connection of Cartan type (3.2), are true:*

$$
\begin{cases}
\Delta^{(1)}_{(i)1|k} - \Delta^{(1)}_{(i)k/1} = p_r^1 R^r_{i1k} - \Delta^{(1)}_{(i)r} T^r_{1k} - \vartheta^{(1)(r)}_{(i)(1)} R^{(1)}_{(r)1k} \\[2mm]
\Delta^{(1)}_{(i)j|k} - \Delta^{(1)}_{(i)k|j} = p_r^1 R^r_{ijk} - \vartheta^{(1)(r)}_{(i)(1)} R^{(1)}_{(r)jk} \\[2mm]
\Delta^{(1)\ |(k)}_{(i)1\ |(1)} - \vartheta^{(1)(k)}_{(i)(1)/1} = p_r^1 P^{r\ \ (k)}_{i1(1)} - \vartheta^{(1)(r)}_{(i)(1)} P^{(1)\ (k)}_{(r)1(1)} \\[2mm]
\Delta^{(1)\ |(k)}_{(i)j\ |(1)} - \vartheta^{(1)(k)}_{(i)(1)|j} = p_r^1 P^{r\ \ (k)}_{ij(1)} - \Delta^{(1)}_{(i)r} C^{r(k)}_{j(1)} - \vartheta^{(1)(r)}_{(i)(1)} P^{(1)\ (k)}_{(r)j(1)} \\[2mm]
\vartheta^{(1)(j)\ |(k)}_{(i)(1)\ |(1)} - \vartheta^{(1)(k)\ |(j)}_{(i)(1)\ |(1)} = p_r^1 S^{r(j)(k)}_{i(1)(1)}.
\end{cases}
\tag{3.4}
$$

3.4   Local Bianchi Identities

It is a well-known fact that the torsion $\mathbf{T}$ and the curvature $\mathbf{R}$ of a connection D on the dual 1-jet space $E^* = J^{1*}(\mathbb{R}, M)$ are connected by the following general *Bianchi identities*:

$$
\sum_{\{X,Y,Z\}} \{(D_X\mathbf{T})(Y, Z) - \mathbf{R}(X, Y)Z + \mathbf{T}(\mathbf{T}(X, Y), Z)\} = 0,
$$

$$
\sum_{\{X,Y,Z\}} (D_X\mathbf{R})(Y, Z, U) + \mathbf{R}(\mathbf{T}(X, Y), Z)U = 0,
$$

where $\sum_{\{X,Y,Z\}}$ means a cyclic sum, for any $X, Y, Z, U \in \mathcal{X}(E^*)$. Obviously, working with a *h*-normal *N*-linear connection of Cartan type and the *N*-adapted basis of *d*-vector fields $(X_A) \subset \mathcal{X}(E^*)$, the above Bianchi identities are locally described by the following equalities:

$$
\sum_{\{A,B,C\}} \{\mathbf{R}^F_{ABC} - \mathbf{T}^F_{AB:C} - \mathbf{T}^G_{AB}\mathbf{T}^F_{CG}\} = 0,
$$

$$
\sum_{\{A,B,C\}} \{\mathbf{R}^F_{DAB:C} + \mathbf{T}^G_{AB}\mathbf{R}^F_{DCG}\} = 0,
\tag{3.5}
$$

where $\mathbf{R}(X_A, X_B)X_C = \mathbf{R}^D_{CBA}X_D$, $\mathbf{T}(X_A, X_B) = \mathbf{T}^D_{BA}X_D$, and "$:_C$" represents one from the local covariant derivatives "$_{/1}$", "$_{|i}$", or "$|^{(i)}_{(1)}$" of the *h*-normal *N*-linear connection of Cartan-type $CD\Gamma(N)$. Consequently, we find

Theorem 3.5 *The following **19 local Bianchi identities** for a h-normal N-linear connection of Cartan type are true on the dual 1-jet space E^*:*

1. $\mathcal{A}_{\{j,k\}}\left\{ C^{l(r)}_{k(1)} R^{(1)}_{(r)1j} + R^l_{j1k} + T^l_{1j|k} \right\} = 0,$

2. $\sum_{\{i,j,k\}}\left\{ C^{l(r)}_{k(1)} R^{(1)}_{(r)ij} - R^l_{ijk} \right\} = 0,$

3. $\mathcal{A}_{\{j,k\}}\left\{ R^{(1)}_{(l)1j|k} + P^{(1)\ (r)}_{(l)k(1)} R^{(1)}_{(r)1j} + R^{(1)}_{(l)kr} T^r_{1j} \right\} = -R^{(1)}_{(l)jk/1} - P^{(1)\ (r)}_{(l)1(1)} R^{(1)}_{(r)jk},$

4. $\sum_{\{i,j,k\}} \left\{ R^{(1)}_{(l)ij|k} + P^{(1)}_{(l)k(1)} {}^{(r)}R^{(1)}_{(r)ij} \right\} = 0,$

5. $T^l_{1k}|^{(p)}_{(1)} - C^{l(p)}_{r(1)} T^r_{1k} + P^{l}_{k1(1)}{}^{(p)} + C^{l(p)}_{k(1)/1} - C^{l(r)}_{k(1)} P^{(1)}_{(r)1(1)}{}^{(p)} + C^{r(p)}_{k(1)} T^l_{1r} = 0,$

6. $\mathcal{A}_{\{j,k\}} \left\{ C^{l(p)}_{j(1)|k} + C^{l(r)}_{k(1)} P^{(1)}_{(r)j(1)}{}^{(p)} + P^{l}_{jk(1)}{}^{(p)} \right\} = 0,$

7. $P^{(1)}_{(l)1(1)|k}{}^{(p)} - P^{(1)}_{(l)k(1)/1}{}^{(p)} + P^{(1)}_{(l)k(1)}{}^{(r)} P^{(1)}_{(r)1(1)}{}^{(p)} - P^{(1)}_{(l)1(1)}{}^{(r)} P^{(1)}_{(r)k(1)}{}^{(p)} =$

$= R^{(1)}_{(l)1k}|^{(p)}_{(1)} + R^{(1)(p)}_{(l)(1)1k} + R^{(1)}_{(l)1r} C^{r(p)}_{k(1)} - T^r_{1k} P^{(1)}_{(l)r(1)}{}^{(p)},$

8. $\mathcal{A}_{\{j,k\}} \left\{ P^{(1)}_{(l)j(1)|k}{}^{(p)} + P^{(1)}_{(l)k(1)}{}^{(r)} P^{(1)}_{(r)j(1)}{}^{(p)} + R^{(1)}_{(l)kr} C^{r(p)}_{j(1)} \right\} = R^{(1)}_{(l)jk}|^{(p)}_{(1)} + R^{(1)(p)}_{(l)(1)jk},$

9. $\mathcal{A}_{\left\{ {(j) \atop (1)} \cdot {(k) \atop (1)} \right\}} \left\{ C^{l(j)}_{i(1)}|^{(k)}_{(1)} + C^{r(k)}_{i(1)} C^{l(j)}_{r(1)} \right\} = S^{l(j)(k)}_{i(1)(1)},$

10. $\mathcal{A}_{\left\{ {(j) \atop (1)} \cdot {(k) \atop (1)} \right\}} \left\{ P^{(1)}_{(l)1(1)}{}^{(j)}|^{(k)}_{(1)} - P^{(1)(j)}_{(l)(1)1(1)}{}^{(k)} \right\} = 0,$

11. $\mathcal{A}_{\left\{ {(j) \atop (1)} \cdot {(k) \atop (1)} \right\}} \left\{ P^{(1)}_{(l)i(1)}{}^{(j)}|^{(k)}_{(1)} - P^{(1)(j)}_{(l)(1)i(1)}{}^{(k)} - C^{r(j)}_{i(1)} P^{(1)}_{(l)r(1)}{}^{(k)} \right\} = 0,$

12. $\sum_{\left\{ {(i) \atop (1)} \cdot {(j) \atop (1)} \cdot {(k) \atop (1)} \right\}} S^{(1)(i)(j)(k)}_{(l)(1)(1)(1)} = 0,$

13. $\mathcal{A}_{\{j,k\}} \left\{ R^l_{p1j|k} + R^{(1)}_{(r)1j} P^{l}_{pk(1)}{}^{(r)} + T^r_{1j} R^l_{pkr} \right\} = -R^l_{pjk/1} - R^{(1)}_{(r)jk} P^{l}_{p1(1)}{}^{(r)},$

14. $\sum_{\{i,j,k\}} \left\{ R^l_{pij|k} + R^{(1)}_{(r)ij} P^{l}_{pk(1)}{}^{(r)} \right\} = 0,$

15. $P^l_{i1(1)|k}{}^{(p)} - P^l_{ik(1)/1}{}^{(p)} + P^{(1)}_{(r)1(1)}{}^{(p)} P^l_{ik(1)}{}^{(r)} - P^{(1)}_{(r)k(1)}{}^{(p)} P^l_{i1(1)}{}^{(r)} =$

$= R^l_{i1k}|^{(p)}_{(1)} + R^{(1)}_{(r)1k} S^{l(p)(r)}_{i(1)(1)} + C^{r(p)}_{k(1)} R^l_{i1r} - T^r_{1k} P^l_{ir(1)}{}^{(p)},$

16. $\mathcal{A}_{\{j,k\}} \left\{ P^l_{ij(1)|k}{}^{(p)} + P^{(1)}_{(r)j(1)}{}^{(p)} P^l_{ik(1)}{}^{(r)} + C^{r(p)}_{j(1)} R^l_{ikr} \right\} = R^l_{ijk}|^{(p)}_{(1)} + R^{(1)}_{(r)jk} S^{l(p)(r)}_{i(1)(1)},$

17. $\mathcal{A}_{\left\{ {(j) \atop (1)} \cdot {(k) \atop (1)} \right\}} \left\{ P^l_{p1(1)}{}^{(j)}|^{(k)}_{(1)} + P^{(1)}_{(r)1(1)}{}^{(j)} S^{l(k)(r)}_{p(1)(1)} \right\} = -S^{l(j)(k)}_{p(1)(1)/1},$

18. $\mathcal{A}_{\left\{ {(j) \atop (1)} \cdot {(k) \atop (1)} \right\}} \left\{ P^l_{pi(1)}{}^{(j)}|^{(k)}_{(1)} + P^{(1)}_{(r)i(1)}{}^{(j)} S^{l(k)(r)}_{p(1)(1)} - C^{r(j)}_{i(1)} P^l_{pr(1)}{}^{(k)} \right\} = -S^{l(j)(k)}_{p(1)(1)|i},$

19. $\sum_{\left\{ {(i) \atop (1)} \cdot {(j) \atop (1)} \cdot {(k) \atop (1)} \right\}} S^{l(i)(j)}_{p(1)(1)}|^{(k)}_{(1)} = 0,$

where, if $\{A, B, C\}$ are indices of type $\left\{ 1, i, {(1) \atop (i)} \right\}$, then $\sum_{\{A,B,C\}}$ represents a cyclic sum, and $\mathcal{A}_{\{A,B\}}$ represents an alternate sum.

Proof Taking into account that the indices $A, B, C, D...$ are of type $\left\{1, i, \overset{(1)}{\underset{(i)}{}}\right\}$, and the torsion $\mathbf{T}^C_{AB}$ and curvature $\mathbf{R}^D_{ABC}$ adapted components are given in Tables 3.1 and 3.2, after laborious local computations, Formula (3.5) implies the required Bianchi identities.   □

Remark 3.4 Some identities from our local Bianchi identities reduce to those 11 Bianchi identities of a *N-linear connection* in the Hamiltonian geometry of cotangent bundles (see Miron et al. [21]).

Distinguished Geometrization of the Time-Dependent Hamiltonians of Momenta

4

Abstract

The aim of this chapter is the geometrization on the 1-jet space $J^{1*}(\mathbb{R}, M)$ of the time-dependent Hamiltonians, in the sense of canonical nonlinear connections, Cartan N-linear connections, d-torsions, and d-curvatures. Some time-dependent Hamiltonian field-like geometrical models (electromagnetic-like and gravitational-like) depending on momenta are also constructed.

4.1 Time-Dependent Hamiltonians of Momenta

The geometrical ideas from this chapter follow the paper Neagu-Balan-Oană [22]. Let us start with a time-dependent Hamiltonian $H : E^* = J^{1*}(\mathbb{R}, M) \to \mathbb{R}$, locally expressed by

$$E^* \ni (t, x^i, p_i^1) \to H(t, x^i, p_i^1) \in \mathbb{R},$$

whose *vertical fundamental metrical d-tensor* is given by

$$G^{(i)(j)}_{(1)(1)} = \frac{1}{2} \frac{\partial^2 H}{\partial p_i^1 \partial p_j^1}.$$

Let $h = (h_{11}(t))$ be a semi-Riemannian metric on the time manifold $\mathbb{R}$, together with a d-tensor $g^{ij}(t, x^k, p_k^1)$ on the dual 1-jet space E^*, which is symmetric, has the rank $n = \dim M$, and has a constant signature.

Definition 4.1 A time-dependent Hamiltonian $H : E^* \to \mathbb{R}$, having the fundamental vertical metrical d-tensor of the form

© The Author(s), under exclusive license to Springer Nature Switzerland AG 2022
M. Neagu and A. Oană, *Dual Jet Geometrization for Time-Dependent Hamiltonians and Applications*, Synthesis Lectures on Mathematics & Statistics, https://doi.org/10.1007/978-3-031-08885-8_4

$$G_{(1)(1)}^{(i)(j)}(t, x^k, p_k^1) = \frac{1}{2}\frac{\partial^2 H}{\partial p_i^1 \partial p_j^1} = h_{11}(t)g^{ij}(t, x^k, p_k^1), \qquad (4.1)$$

is called a **Kronecker h-regular time-dependent Hamiltonian function**.

Remark 4.1 If we take the temporal metric as a Riemannian one on $\mathbb{R}$ of the form $h_{11}(t) = e^{\sigma(t)}$, we deduce that the vertical metrical d-tensor (4.1) is in fact a temporal conformal deformation of the metrical tensor $g^{ij}(t, x^k, p_k^1)$.

In this geometrical context, we can introduce the following notion.

Definition 4.2 A pair of mathematical objects $H^n = (E^*, H)$, consisting of the dual 1-jet space $E^* = J^{1*}(\mathbb{R}, M)$ and a Kronecker h-regular time-dependent Hamiltonian $H : E^* \to \mathbb{R}$, is called a **time-dependent Hamilton space**.

Example

If $h_{11}(t)$ (respectively, $\varphi_{ij}(x)$) is a semi-Riemannian metric on the time (respectively, spatial) manifold $\mathbb{R}$ (respectively, M) having the physical meaning of **gravitational potentials**, and m, c, and e are the well-known constants from Theoretical Physics representing the **mass of the test body**, **speed of light**, and **electric charge**, then let us consider the Kronecker h-regular time-dependent Hamiltonian $H_1 : E^* \to \mathbb{R}$, defined by

$$H_1 = \frac{1}{4mc}h_{11}(t)\varphi^{ij}(x)p_i^1 p_j^1 - \frac{e}{m^2c}A_{(1)}^{(i)}(x)p_i^1 + \frac{e^2}{m^3c}F(t, x) - \mathsf{P}(t, x), \qquad (4.2)$$

where $A_{(1)}^{(i)}(x)$ is a d-tensor on E^* having the physical meaning of a **potential d-tensor of an electromagnetic field**, $\mathsf{P}(t, x)$ is a **potential function**, and the function $F(t, x)$ is given by

$$F(t, x) = h^{11}(t)\varphi_{ij}(x)A_{(1)}^{(i)}(x)A_{(1)}^{(j)}(x).$$

Then, the Hamilton space $\mathcal{EDH}^n = (E^*, H_1)$ defined by the time-dependent Hamiltonian (4.2) is called the **time-dependent Hamilton space of electrodynamics of autonomous type**. This is because, in the particular case of the metric $h = \delta = 1$, we recover the classical Hamilton space of electrodynamics studied in the monograph [23]. The non-dynamical character (i.e., the independence on the temporal coordinate t) of the spatial gravitational potentials $\varphi_{ij}(x)$ motivated us to use the term **"autonomous"**. ◀

Example

More general, if we take on E^* a symmetric d-tensor field $g_{ij}(t, x)$ having the rank n and a constant signature, we can define the Kronecker h-regular time-dependent Hamiltonian

$H_2 : E^* \to \mathbb{R}$, by putting

$$H_2 = h_{11}(t)g^{ij}(t, x)p_i^1 p_j^1 + U_{(1)}^{(i)}(t, x)p_i^1 + \mathcal{F}(t, x), \tag{4.3}$$

where $U_{(1)}^{(i)}(t, x)$ is a d-tensor field on E^* and $\mathcal{F}(t, x)$ is a function on E^*. Then, the Hamilton space $\mathcal{NEDH}^n = (E^*, H_2)$, defined by the affine quadratic time-dependent Hamiltonian (4.3), is called the **non-autonomous time-dependent Hamilton space of electrodynamics**. The dynamical character (i.e., the dependence on the temporal coordinate t) of the gravitational potentials $g_{ij}(t, x)$ motivated us to use the syntagma "**non-autonomous**". ◀

4.2 Canonical Nonlinear Connections on H^n-Spaces

In the sequel, following Miron's geometrical ideas from [23, 24], we will prove that any Kronecker h-regular time-dependent Hamiltonian H produces a natural nonlinear connection on the dual 1-jet bundle E^*, which is determined only by H. At the beginning, let us consider a Kronecker h-regular time-dependent Hamiltonian H, whose fundamental vertical metrical d-tensor is given by (4.1). Also, let us introduce the *generalized Christoffel symbols* of the inverse spatial metrical d-tensor $g_{ij}(t, x^k, p_k^1)$, where $g^{ij}(t, x^k, p_k^1) = h^{11}(t)G_{(1)(1)}^{(i)(j)}(t, x^k, p_k^1)$, via the formulas

$$\Gamma_{ij}^k = \frac{g^{kl}}{2}\left(\frac{\partial g_{li}}{\partial x^j} + \frac{\partial g_{lj}}{\partial x^i} - \frac{\partial g_{ij}}{\partial x^l}\right).$$

In this context, using above notations, we can give the following result.

Theorem 4.1 *The pair of local functions* $N = \left(\underset{1}{N}_{(i)1}^{(1)}, \underset{2}{N}_{(i)j}^{(1)}\right)$ *on* E^**, where*

$$\underset{1}{N}_{(i)1}^{(1)} = H_{11}^1 p_i^1 = (h^{11}/2)(dh_{11}/dt)p_i^1,$$

$$\underset{2}{N}_{(i)j}^{(1)} = \frac{h^{11}}{4}\left[\frac{\partial g_{ij}}{\partial x^k}\frac{\partial H}{\partial p_k^1} - \frac{\partial g_{ij}}{\partial p_k^1}\frac{\partial H}{\partial x^k} + g_{ik}\frac{\partial^2 H}{\partial x^j \partial p_k^1} + g_{jk}\frac{\partial^2 H}{\partial x^i \partial p_k^1}\right] \tag{4.4}$$

represents a nonlinear connection on E^**, which is called the **canonical nonlinear connection of the time-dependent Hamilton space** $H^n = (E^*, H)$.*

Proof Taking into account the transformation rule of the Christoffel symbol H_{11}^1 of the temporal semi-Riemannian metric h_{11}, by direct local computations, we deduce that the temporal components $\underset{1}{N}_{(i)1}^{(1)}$ from (4.4) verify the transformation rules of a temporal nonlinear connection (1.8).

The spatial components from (4.4) become (except the multiplication factor h^{11}) exactly the canonical nonlinear connection from the classical Hamilton geometry (see [23, p. 127] or [24]).   □

4.3   Cartan Canonical Connection in H^n-Spaces

Let $H^n = (E^* = J^{1*}(\mathbb{R}, M), H)$ be a time-dependent Hamilton space, whose fundamental vertical metrical d-tensor is given by (4.1). Let

$$N = \left(\underset{1}{N}_{(i)1}^{(1)}, \; \underset{2}{N}_{(i)j}^{(1)} \right)$$

be the canonical nonlinear connection of the time-dependent Hamilton space H^n, given by (4.4).

Theorem 4.2 (the   Cartan   canonical N-linear   connection)   *On   the   time-dependent Hamilton space $H^n = (E^*, H)$, endowed with the canonical nonlinear connection (4.4), there exists an unique h-normal N-linear connection*

$$C\Gamma(N) = \left(H_{11}^1, A_{j1}^i, H_{jk}^i, C_{j(1)}^{i(k)} \right),   \tag{4.5}$$

having the following metrical properties:

(i) $g_{ij|k} = 0, \quad g^{ij}|_{(1)}^{(k)} = 0;$

(ii) $A_{j1}^i = \dfrac{g^{il}}{2} \dfrac{\delta g_{lj}}{\delta t}, \quad H_{jk}^i = H_{kj}^i, \quad C_{j(1)}^{i(k)} = C_{j(1)}^{k(i)};$

where "$_{/1}$", "$_{|k}$", and "$|_{(1)}^{(k)}$" represent the local covariant derivatives induced by the h-normal N-linear connection $C\Gamma(N)$.

Proof Let $C\Gamma(N) = \left(H_{11}^1, A_{j1}^i, H_{jk}^i, C_{j(1)}^{i(k)} \right)$ be a h-normal N-linear connection, whose local coefficients are defined by the following relations:

$$A_{11}^1 = H_{11}^1 = \frac{h^{11}}{2} \frac{dh_{11}}{dt}, \qquad A_{j1}^i = \frac{g^{il}}{2} \frac{\delta g_{lj}}{\delta t},$$

$$H_{jk}^i = \frac{g^{ir}}{2} \left(\frac{\delta g_{jr}}{\delta x^k} + \frac{\delta g_{kr}}{\delta x^j} - \frac{\delta g_{jk}}{\delta x^r} \right), \quad C_{i(1)}^{j(k)} = -\frac{g_{ir}}{2} \left(\frac{\partial g^{jr}}{\partial p_k^1} + \frac{\partial g^{kr}}{\partial p_j^1} - \frac{\partial g^{jk}}{\partial p_r^1} \right).$$

Taking into account the local expressions of the local covariant derivatives induced by the h-normal N-linear connection $C\Gamma(N)$, by local computations, we infer that $C\Gamma(N)$ satisfies conditions (i) and (ii).

Conversely, let us consider a *h*-normal *N*-linear connection

$$\tilde{C}\Gamma(N) = \left(\tilde{A}^1_{11}, \tilde{A}^i_{j1}, \tilde{H}^i_{jk}, \tilde{C}^{i(k)}_{j(1)} \right),$$

which satisfies conditions (i) and (ii). It follows that we have

$$\tilde{A}^1_{11} = H^1_{11}, \quad \tilde{A}^i_{j1} = \frac{g^{il}}{2} \frac{\delta g_{lj}}{\delta t}.$$

Moreover, the metrical condition $g_{ij|k} = 0$ is equivalent with

$$\frac{\delta g_{ij}}{\delta x^k} = g_{rj} \tilde{H}^r_{ik} + g_{ir} \tilde{H}^r_{jk}.$$

Applying now a Christoffel process to indices $\{i, j, k\}$, we get

$$\tilde{H}^i_{jk} = \frac{g^{ir}}{2} \left(\frac{\delta g_{jr}}{\delta x^k} + \frac{\delta g_{kr}}{\delta x^j} - \frac{\delta g_{jk}}{\delta x^r} \right).$$

By analogy, using the relations $C^{i(k)}_{j(1)} = C^{k(i)}_{j(1)}$ and $g^{ij}|^{(k)}_{(1)} = 0$, together with a Christoffel process applied to indices $\{i, j, k\}$, we find

$$\tilde{C}^{j(k)}_{i(1)} = -\frac{g_{ir}}{2} \left(\frac{\partial g^{jr}}{\partial p^1_k} + \frac{\partial g^{kr}}{\partial p^1_j} - \frac{\partial g^{jk}}{\partial p^1_r} \right).$$

In conclusion, the uniqueness of the *Cartan canonical connection* $C\Gamma(N)$ on the dual 1-jet space $E^* = J^{1*}(\mathbb{R}, M)$ is obvious.   □

Remark 4.2 The Cartan canonical connection $C\Gamma(N)$ of the time-dependent Hamilton space H^n also verifies the metrical properties

$$h_{11/1} = h_{11|k} = h_{11}|^{(k)}_{(1)} = 0, \quad g_{ij/1} = 0.$$

4.4   *d*-Torsions and *d*-Curvatures

By applying the formulas of the local *d*-torsions and *d*-curvatures of a *h*-normal *N*-linear connection $D\Gamma(N)$ (see Tables 3.1 and 3.2) to the Cartan canonical connection $C\Gamma(N)$, we get the following important geometrical results.

Theorem 4.3 *The torsion tensor* **T** *of the Cartan canonical connection* $C\Gamma(N)$ *of the time-dependent Hamilton space* H^n *is determined by the local d-components (see also Table 4.1):*

$$T^r_{1j} = -A^r_{j1}, \quad P^{r(j)}_{i(1)} = C^{r(j)}_{i(1)},$$

Table 4.1 d-torsions of Cartan canonical connection

	$h_{\mathbb{R}}$	h_M	v
$h_{\mathbb{R}}h_{\mathbb{R}}$	0	0	0
$h_M h_{\mathbb{R}}$	0	T^r_{1j}	$R^{(1)}_{(r)1j}$
$vh_{\mathbb{R}}$	0	0	$P^{(1)\ (j)}_{(r)1(1)}$
$h_M h_M$	0	0	$R^{(1)}_{(r)ij}$
vh_M	0	$P^{r(j)}_{i(1)}$	$P^{(1)\ (j)}_{(r)i(1)}$
vv	0	0	0

$$P^{(1)\ (j)}_{(r)1(1)} = \frac{\partial N^{(1)}_{(r)1}}{\partial p^1_j} + A^j_{r1} - \delta^j_r H^1_{11}, \quad P^{(1)\ (j)}_{(r)i(1)} = \frac{\partial N^{(1)}_{2\ (r)i}}{\partial p^1_j} + H^j_{ri},$$

$$R^{(1)}_{(r)1j} = \frac{\delta N^{(1)}_{(r)1}}{\delta x^j} - \frac{\delta N^{(1)}_{2\ (r)j}}{\delta t}, \quad R^{(1)}_{(r)ij} = \frac{\delta N^{(1)}_{2\ (r)i}}{\delta x^j} - \frac{\delta N^{(1)}_{2\ (r)j}}{\delta x^i}.$$

Theorem 4.4 *The curvature tensor* **R** *of the Cartan canonical connection* $C\Gamma(N)$ *of the time-dependent Hamilton space* H^n *is determined by the following adapted local curvature d-tensors (see also Table 4.2):*

$$R^l_{i1k} = \frac{\delta A^l_{i1}}{\delta x^k} - \frac{\delta H^l_{ik}}{\delta t} + A^r_{i1}H^l_{rk} - H^r_{ik}A^l_{r1} + C^{l(r)}_{i(1)}R^{(1)}_{(r)1k},$$

$$R^l_{ijk} = \frac{\delta H^l_{ij}}{\delta x^k} - \frac{\delta H^l_{ik}}{\delta x^j} + H^r_{ij}H^l_{rk} - H^r_{ik}H^l_{rj} + C^{l(r)}_{i(1)}R^{(1)}_{(r)jk},$$

$$P^{l\ (k)}_{i1(1)} = \frac{\partial A^l_{i1}}{\partial p^1_k} - C^{l(k)}_{i(1)/1} + C^{l(r)}_{i(1)}P^{(1)\ (k)}_{(r)1(1)},$$

$$P^{l\ (k)}_{ij(1)} = \frac{\partial H^l_{ij}}{\partial p^1_k} - C^{l(k)}_{i(1)|j} + C^{l(r)}_{i(1)}P^{(1)\ (k)}_{(r)j(1)},$$

$$S^{l(j)(k)}_{i(1)(1)} = \frac{\partial C^{l(j)}_{i(1)}}{\partial p^1_k} - \frac{\partial C^{l(k)}_{i(1)}}{\partial p^1_j} + C^{r(j)}_{i(1)}C^{l(k)}_{r(1)} - C^{r(k)}_{i(1)}C^{l(j)}_{r(1)}.$$

Table 4.2 d-curvatures of Cartan canonical connection

	$h_\mathbb{R}$	h_M	v
$h_\mathbb{R}h_\mathbb{R}$	0	0	0
$h_M h_\mathbb{R}$	0	R^l_{i1k}	$-R^{(1)(l)}_{(i)(1)1k} = -R^l_{i1k}$
$vh_\mathbb{R}$	0	$P^{l\ (k)}_{i1(1)}$	$-P^{(1)(l)\ (k)}_{(i)(1)1(1)} = -P^{l\ (k)}_{i1(1)}$
$h_M h_M$	0	R^l_{ijk}	$-R^{(1)(l)}_{(i)(1)jk} = -R^l_{ijk}$
vh_M	0	$P^{l\ (k)}_{ij(1)}$	$-P^{(1)(l)\ (k)}_{(i)(1)j(1)} = -P^{l\ (k)}_{ij(1)}$
vv	0	$S^{l(j)(k)}_{i(1)(1)}$	$-S^{(1)(l)(j)(k)}_{(i)(1)(1)(1)} = -S^{l(j)(k)}_{i(1)(1)}$

4.5 Momentum Field-Like Geometrical Models

In what follows, we create a large geometrical framework on the dual 1-jet space E^* for a time-dependent Hamiltonian approach of the electromagnetic and gravitational physical fields. Our geometric-physical construction is achieved starting only from a given time-dependent Hamiltonian function H, which naturally produces a canonical nonlinear connection N, a canonical Cartan N-linear connection $C\Gamma(N)$, and their corresponding local d-torsions and d-curvatures. In this context, we construct some geometrical time-dependent Hamiltonian electromagnetic-like and gravitational-like field theories, governed by some natural geometrical momentum Maxwell-like and Einstein-like equations.

4.5.1 Geometrical Momentum Maxwell-Like Equations

Let $H^n = (E^*, H)$ be a time-dependent Hamilton space, endowed with its canonical nonlinear connection (4.4), which produces the adapted vertical distinguished 1-forms

$$\delta p^1_i = dp^1_i + \underset{1}{N}^{(1)}_{(i)1}dt + \underset{2}{N}^{(1)}_{(i)j}dx^j.$$

Consider that $C\Gamma(N)$ is the Cartan canonical linear connection of the space H^n locally defined by (4.5). Let us also take the canonical Liouville-Hamilton d-tensor field of momenta $\mathbb{C}^* = p^1_i(\partial/\partial p^1_i)$, together with the fundamental vertical metrical d-tensor (4.1). All these geometrical objects allow us to define the *metrical deflection d-tensors*

$$\Delta^{(i)}_{(1)1} = p^{(i)}_{(1)/1}, \quad \Delta^{(i)}_{(1)j} = p^{(i)}_{(1)|j}, \quad \vartheta^{(i)(j)}_{(1)(1)} = p^{(i)}_{(1)}|^{(j)}_{(1)},$$

where $p^{(i)}_{(1)} = G^{(i)(k)}_{(1)(1)} p^1_k$ and "$/_1$", "$_{|j}$", and "$|^{(j)}_{(1)}$" are the local covariant derivatives induced by the Cartan connection $C\Gamma(N)$. Taking into account the form of the local covariant derivatives of the Cartan canonical connection $C\Gamma(N)$, by direct computations, we get

Proposition 4.1 *The metrical deflection d-tensors of the time-dependent Hamilton space H^n are given by*

$$\Delta^{(i)}_{(1)1} = -h_{11} g^{ik} A^r_{k1} p^1_r, \quad \Delta^{(i)}_{(1)j} = h_{11} g^{ik} \left[-N^{(1)}_{(k)j} - H^r_{kj} p^1_r \right],$$

$$\vartheta^{(i)(j)}_{(1)(1)} = h_{11} g^{ij} - h_{11} g^{ik} C^{r(j)}_{k(1)} p^1_r.$$

In order to construct our time-dependent Hamiltonian theory of electromagnetism, we introduce the following geometric-physical notion.

Definition 4.3 The distinguished 2-form on the 1-jet space E^*, defined by

$$\mathbb{F} = F^{(i)}_{(1)j} \delta p^1_i \wedge dx^j + f^{(i)(j)}_{(1)(1)} \delta p^1_i \wedge \delta p^1_j, \tag{4.6}$$

where

$$F^{(i)}_{(1)j} = \frac{1}{2} \left[\Delta^{(i)}_{(1)j} - \Delta^{(j)}_{(1)i} \right], \quad f^{(i)(j)}_{(1)(1)} = \frac{1}{2} \left[\vartheta^{(i)(j)}_{(1)(1)} - \vartheta^{(j)(i)}_{(1)(1)} \right], \tag{4.7}$$

is called the **electromagnetic field of the time-dependent Hamilton space H^n** or **momentum electromagnetic field**.

By a straightforward calculation, we infer.

Proposition 4.2 *The local components $F^{(i)}_{(1)j}$ and $f^{(i)(j)}_{(1)(1)}$ of the electromagnetic field $\mathbb{F}$, associated with the Hamilton space H^n, have the following expressions:*

$$F^{(i)}_{(1)j} = \frac{h^{11}}{2} \left[g^{jk} N^{(1)}_{(k)i} - g^{ik} N^{(1)}_{(k)j} + \left(g^{jk} H^r_{ki} - g^{ik} H^r_{kj} \right) p^1_r \right], \quad f^{(i)(j)}_{(1)(1)} = 0.$$

The main result of our abstract geometrical Hamilton time-dependent electromagnetism of momenta is

Theorem 4.5 *The electromagnetic components $F^{(i)}_{(1)j}$ of the space H^n are governed by the following **geometrical Maxwell-like equations**:*

$$\begin{cases} F^{(i)}_{(1)k/1} = \dfrac{1}{2}\mathcal{A}_{\{i,k\}} \left\{ \Delta^{(i)}_{(1)1|k} + \Delta^{(i)}_{(1)r} T^r_{1k} + \vartheta^{(i)(r)}_{(1)(1)} R^{(1)}_{(r)1k} + R^i_{r1k} p^{(r)}_{(1)} \right\} \\[2mm] \sum_{\{i,j,k\}} F^{(i)}_{(1)j|k} = -\dfrac{1}{2} \sum_{\{i,j,k\}} \left\{ \vartheta^{(i)(r)}_{(1)(1)} R^{(1)}_{(r)jk} + R^i_{rjk} p^{(r)}_{(1)} \right\} \\[2mm] F^{(i)}_{(1)j}|^{(k)}_{(1)} = \dfrac{1}{2}\mathcal{A}_{\{i,j\}} \left\{ \vartheta^{(i)(k)}_{(1)(1)|j} - P^{i\ (k)}_{rj(1)} p^{(r)}_{(1)} - \Delta^{(i)}_{(1)r} C^{r(k)}_{j(1)} - \vartheta^{(i)(r)}_{(1)(1)} P^{(1)\ (k)}_{(r)j(1)} \right\}, \end{cases}$$

where $\mathcal{A}_{\{i,j\}}$ means an alternate sum and $\sum_{\{i,j,k\}}$ means a cyclic sum.

Proof The general Ricci identities applied to the metric g^{ij} give us the equalities (see also [25]):

$$g^{ir} R^j_{r1k} + g^{jr} R^i_{r1k} = 0, \qquad g^{ir} R^j_{rkl} + g^{jr} R^i_{rkl} = 0, \tag{4.8}$$
$$g^{ir} P^{j\ (l)}_{rk(1)} + g^{jr} P^{i\ (l)}_{rk(1)} = 0.$$

Let us consider now the non-metrical deflection d-tensor identities (3.4):

(d_1) $\Delta^{(1)}_{(p)1|k} - \Delta^{(1)}_{(p)k/1} = p^1_r R^r_{p1k} - \Delta^{(1)}_{(p)r} T^r_{1k} - \vartheta^{(1)(r)}_{(p)(1)} R^{(1)}_{(r)1k},$

(d_2) $\Delta^{(1)}_{(p)j|k} - \Delta^{(1)}_{(p)k|j} = p^1_r R^r_{pjk} - \vartheta^{(1)(r)}_{(p)(1)} R^{(1)}_{(r)jk},$

(d_3) $\Delta^{(1)}_{(p)j}|^{(k)}_{(1)} - \vartheta^{(1)(k)}_{(p)(1)|j} = p^1_r P^{r\ (k)}_{pj(1)} - \Delta^{(1)}_{(p)r} C^{r(k)}_{j(1)} - \vartheta^{(1)(r)}_{(p)(1)} P^{(1)\ (k)}_{(r)j(1)},$

where

$$\Delta^{(1)}_{(i)1} = p^1_{i/1}, \qquad \Delta^{(1)}_{(i)j} = p^1_{i|j}, \qquad \vartheta^{(1)(j)}_{(i)(1)} = p^1_i|^{(j)}_{(1)}.$$

Contracting the above deflection d-tensor identities with the fundamental vertical metrical d-tensor $G^{(i)(p)}_{(1)(1)}$, and using equalities (4.8), we obtain the following *metrical deflection d-tensor identities:*

(d'_1) $\Delta^{(i)}_{(1)1|k} - \Delta^{(i)}_{(1)k/1} = -p^{(r)}_{(1)} R^i_{r1k} - \Delta^{(i)}_{(1)r} T^r_{1k} - \vartheta^{(i)(r)}_{(1)(1)} R^{(1)}_{(r)1k},$

(d'_2) $\Delta^{(i)}_{(1)j|k} - \Delta^{(i)}_{(1)k|j} = -p^{(r)}_{(1)} R^i_{rjk} - \vartheta^{(i)(r)}_{(1)(1)} R^{(1)}_{(r)jk},$

(d'_3) $\Delta^{(i)}_{(1)j}|^{(k)}_{(1)} - \vartheta^{(i)(k)}_{(1)(1)|j} = -p^{(r)}_{(1)} P^{i\ (k)}_{rj(1)} - \Delta^{(i)}_{(1)r} C^{r(k)}_{j(1)} - \vartheta^{(i)(r)}_{(1)(1)} P^{(1)\ (k)}_{(r)j(1)}.$

To obtain the first (respectively, the third) geometrical Maxwell-like equation, we permute the indices i and k in the identity (d'_1) (respectively, the indices i and j in the identity (d'_3)), and we subtract this new identity from the initial one. Moreover, doing a cyclic sum by indices $\{i, j, k\}$ in the identity (d'_2), it follows the second geometrical Maxwell-like equation. $\square$

4.5.2   Geometrical Momentum Einstein-Like Equations

On a time-dependent Hamilton space $H^n = (E^*, H)$, via its fundamental vertical metrical d-tensor given by (4.1) and its canonical nonlinear connection (4.4), we construct a corresponding *momentum time-dependent gravitational h-potential*, by taking

$$\mathbb{G} = h_{11}dt \otimes dt + g_{ij}dx^i \otimes dx^j + h_{11}g^{ij}\delta p_i^1 \otimes \delta p_j^1.$$

At the same time, let us consider that $C\Gamma(N)$, which is given by (4.5), is the Cartan canonical connection of the time-dependent Hamilton space H^n. We postulate that the geometrical momentum Einstein-like equations, which govern the time-dependent gravitational h-potential $\mathbb{G}$ of the Hamilton space H^n, are the abstract geometrical Einstein equations associated with the Cartan canonical connection $C\Gamma(N)$ and to the adapted metric $\mathbb{G}$ on E^*, namely,

$$\text{Ric}(C\Gamma(N)) - \frac{\text{Sc}(C\Gamma(N))}{2}\mathbb{G} = \mathcal{K}\mathbb{T}, \qquad (4.9)$$

where $\text{Ric}(C\Gamma(N))$ represents the distinguished *Ricci tensor* of the Cartan connection, $\text{Sc}(C\Gamma(N))$ is the *scalar curvature*, $\mathcal{K}$ is the *Einstein constant*, and $\mathbb{T}$ is an intrinsic d-tensor of matter, which is called the *momentum stress-energy d-tensor*.

In the adapted basis of vector fields $(X_A) = (\delta/\delta t, \delta/\delta x^i, \partial/\partial p_i^1)$, produced by the canonical nonlinear connection (4.4), the curvature tensor $\mathbf{R}$ of the Cartan canonical connection $C\Gamma(N)$ is locally expressed by $\mathbf{R}(X_C, X_B)X_A = \mathbf{R}_{ABC}^D X_D$. It follows that we have $R_{AB} = \text{Ric}(X_A, X_B) = \mathbf{R}_{ABD}^D$, and $\text{Sc}(C\Gamma) = G^{AB}R_{AB}$, where

$$G^{AB} = \begin{cases} h^{11}, & \text{for } A = 1, \ B = 1 \\ g^{ij}, & \text{for } A = i, \ B = j \\ h^{11}g_{ij}, & \text{for } A = \overset{(1)}{(i)}, \ B = \overset{(1)}{(j)} \\ 0, & \text{otherwise.} \end{cases} \qquad (4.10)$$

Taking into account, on the one hand, the form of the inverse metrical d-tensor $\mathbf{G}^* = (G_{AB})$ of the time-dependent Hamilton space H^n, and, on the other hand, the expressions of the local curvature d-tensors attached to the Cartan canonical connection $C\Gamma(N)$, by direct computations, we get

Proposition 4.3 *The Ricci d-tensor* $\text{Ric}(C\Gamma(N))$ *of the Cartan canonical connection* $C\Gamma(N)$ *of the time-dependent Hamilton space* H^n *is determined by the following adapted components:*

$$R_{11} := H_{11} = 0, \qquad\qquad R_{1i} = R_{1i1}^1 = 0,$$

$$R_{1(1)}^{(i)} = -P_{1(1)1}^{1(i)} = 0, \qquad R_{i1} = R_{i1r}^r, \qquad R_{ij} = R_{ijr}^r,$$

$$R_{(1)1}^{(i)} := -P_{(1)1}^{(i)} = -P_{r1(1)}^{i\ (r)}, \qquad R_{i(1)}^{(j)} := -P_{i(1)}^{(j)} = -P_{ir(1)}^{r\ (j)},$$

$$R_{(1)(1)}^{(i)(j)} := -S_{(1)(1)}^{(i)(j)} = -S_{r(1)(1)}^{i(j)(r)}, \ R_{(1)j}^{(i)} := -P_{(1)j}^{(i)} = -P_{rj(1)}^{i\ (r)}.$$

Using the notations $R = g^{ij}R_{ij}$ and $S = h^{11}g_{ij}S_{(1)(1)}^{(i)(j)}$, we find

Corollary 4.1 *The scalar curvature* $\mathrm{Sc}(C\Gamma(N))$ *of the Cartan canonical connection* $C\Gamma(N)$ *of the space* H^n *is*

$$\mathrm{Sc}(C\Gamma(N)) = R - S.$$

In this context, the main result of the Hamilton geometrical momentum gravitational theory is

Theorem 4.6 *The* **geometrical Einstein-like equations**, *which govern the gravitational h-potential* $\mathbb{G}$ *of the time-dependent Hamilton space* H^n, *have the following adapted local form:*

$$\begin{cases} -\dfrac{R-S}{2}h_{11} = \mathcal{K}\mathbb{T}_{11} \\[2mm] R_{ij} - \dfrac{R-S}{2}g_{ij} = \mathcal{K}\mathbb{T}_{ij} \\[2mm] -S^{(i)(j)}_{(1)(1)} - \dfrac{R-S}{2}h_{11}g^{ij} = \mathcal{K}\mathbb{T}^{(i)(j)}_{(1)(1)} \\[2mm] 0 = \mathbb{T}_{1i}, \quad R_{i1} = \mathcal{K}\mathbb{T}_{i1}, \quad -P^{(i)}_{(1)1} = \mathcal{K}\mathbb{T}^{(i)}_{(1)1}, \\[2mm] 0 = \mathbb{T}^{(i)}_{1(1)}, \quad -P^{(j)}_{i(1)} = \mathcal{K}\mathbb{T}^{(j)}_{i(1)}, \quad -P^{(i)}_{(1)j} = \mathcal{K}\,\mathbb{T}^{(i)}_{(1)j}, \end{cases} \qquad (4.11)$$

where $\mathbb{T}_{AB}$, $A, B \in \left\{1, i, {}^{(i)}_{(1)}\right\}$ *represent the adapted components of the momentum stress-energy d-tensor of matter* $\mathbb{T}$.

From a theoretical-physics point of view, it is well known that in the classical Riemannian theory of gravity, the stress-energy d-tensor of matter must verify some conservation laws. By a natural extension of the Riemannian conservation laws, in our geometrical Hamiltonian context, we postulate the following *momentum conservation laws* of the stress-energy d-tensor $\mathbb{T}$:

$$\mathbb{T}^B_{A|B} = 0, \quad \forall A \in \left\{1, i, {}^{(1)}_{(i)}\right\},$$

where $\mathbb{T}^B_A = G^{BD}\mathbb{T}_{DA}$. Consequently, by direct computations, we find

Theorem 4.7 *The* **momentum conservation laws** *of the time-dependent Hamilton space* H^n *are given by the following equations:*

$$\begin{cases} \left[\dfrac{R-S}{2}\right]_{/1} = R^r_{1|r} - P^{(1)}_{(r)1}|^{(r)}_{(1)} \\[3mm] \left[R^r_j - \dfrac{R-S}{2}\delta^r_j\right]_{|r} = P^{(1)}_{(r)j}|^{(r)}_{(1)} \\[3mm] \left[S^{(1)(j)}_{(r)(1)} + \dfrac{R-S}{2}\delta^j_r\right]|^{(r)}_{(1)} = -P^{r(j)}_{(1)|r}, \end{cases} \qquad (4.12)$$

where

$$R_1^i = g^{iq} R_{q1}, \qquad P_{(i)1}^{(1)} = h^{11} g_{iq} P_{(1)1}^{(q)}, \; R_j^i = g^{iq} R_{qj},$$

$$P_{(i)j}^{(1)} = h^{11} g_{iq} P_{(1)j}^{(q)}, \; P_{(1)}^{i(j)} = g^{iq} P_{q(1)}^{(j)} \quad S_{(i)(1)}^{(1)(j)} = h^{11} g_{iq} S_{(1)(1)}^{(q)(j)}.$$

Part II
Applications to Dynamical Systems, Economy and Theoretical Physics

The Time-Dependent Hamiltonian of the Least Squares Variational Method

5

Abstract

An application of the dual jet Hamilton geometry to the time-dependent Hamiltonian of the least squares variational method applied to dynamical systems (method that was initiated by C. Udrişte) is studied in this chapter. The corresponding geometrical results for a dynamical system coming from economy are also given.

5.1 Hamiltonian d-Torsions and d-Curvatures of a Dynamical System

The results of this chapter follow the paper Neagu-Balan-Oană [26]. Let us consider a non-autonomous dynamical system, given by

$$\frac{dx^i}{dt} = X^{(i)}_{(1)}(t, x^k(t)),\tag{5.1}$$

where $X^{(i)}_{(1)}(t, x)$ is a d-tensor on $\mathbb{R} \times M$, whose solutions are the global minimum points of the *least squares Lagrangian* (see Udrişte [27] and Neagu-Udrişte [28])

$$L = h^{11}(t)\varphi_{ij}(x)\left(y^i_1 - X^{(i)}_{(1)}\right)\left(y^j_1 - X^{(j)}_{(1)}\right)\tag{5.2}$$

$$= h^{11}\varphi_{ij}y^i_1 y^j_1 - 2h^{11}\varphi_{ij}X^{(i)}_{(1)}y^j_1 + h^{11}\varphi_{ij}X^{(i)}_{(1)}X^{(j)}_{(1)},$$

where $y^i_1 = dx^i/dt$ and $\varphi_{ij}(x)$ is a Riemannian metric on the spatial manifold M, whose Christoffel symbols are $\gamma^i_{jk}(x)$. The Hamiltonian associated with the Lagrangian (5.2) is given by

$$H = \frac{h_{11}\varphi^{ij}}{4}p^1_i p^1_j + X^{(k)}_{(1)}p^1_k,\tag{5.3}$$

© The Author(s), under exclusive license to Springer Nature Switzerland AG 2022
M. Neagu and A. Oană, *Dual Jet Geometrization for Time-Dependent Hamiltonians and Applications*, Synthesis Lectures on Mathematics & Statistics,
https://doi.org/10.1007/978-3-031-08885-8_5

where $p_k^l = \partial L / \partial y_1^k$ and $H = p_k^1 y_1^k - L$. This is called the *least squares Hamiltonian* associated with the dynamical system (5.1).

But, the differential geometry of such time-dependent Hamiltonians was developed in Part I of this monograph. Consequently, we can construct the distinguished geometry of the Hamiltonian (5.3), in the sense of canonical nonlinear connections, Cartan N-linear connections, d-torsions and d-curvatures or momentum electromagnetic-like 2-form. For instance, by direct computations, the canonical nonlinear connection N of the time-dependent Hamiltonian function (5.3) has the components (see also Formula (4.4))

$$\underset{1}{N}{}^{(1)}_{(i)1} = H_{11}^1 p_i^1, \quad \underset{2}{N}{}^{(1)}_{(i)j} = -\gamma_{ij}^k p_k^1 + h^{11}\left(X_{i1\cdot j} + X_{j1\cdot i}\right), \tag{5.4}$$

where $X_{i1} = \varphi_{ik} X_{(1)}^{(k)}$ and

$$X_{k1\cdot r} = \frac{\partial X_{k1}}{\partial x^r} - X_{s1}\gamma_{kr}^s.$$

Moreover, the coefficients of the generalized Cartan canonical connection $C\Gamma(N)$ of the least squares Hamiltonian function (5.3) reduce to

$$A_{11}^1 = H_{11}^1, \quad A_{j1}^i = 0, \quad H_{jk}^i = \gamma_{jk}^i, \quad C_{j(1)}^{i(k)} = 0. \tag{5.5}$$

Remark 5.1 If we have $h_{11} = 1$ and $\varphi_{ij} = \delta_{ij}$, we find the coefficients of the canonical nonlinear connection produced by the least squares Hamiltonian function (5.3) as being the following:

$$\underset{1}{N}{}^{(1)}_{(i)1} = 0, \quad \underset{2}{N}{}^{(1)}_{(i)j} = \frac{\partial X_{(1)}^{(i)}}{\partial x^j} + \frac{\partial X_{(1)}^{(j)}}{\partial x^i}.$$

Moreover, all coefficients of the Cartan canonical connection $C\Gamma(N)$ of the least squares Hamiltonian function (5.3) are zero.

By applying the formulas that give the local d-torsions and d-curvatures of the generalized Cartan canonical connection $C\Gamma(N)$, we obtain the following important geometrical results.

Theorem 5.1 *The torsion tensor* $\mathbf{T}$ *of the generalized Cartan canonical connection* $C\Gamma(N)$ *associated with the least squares Hamiltonian (5.3) is determined by the local d-components*

$$\underset{1}{R}{}^{(1)}_{(r)1j} = -\frac{\partial \underset{2}{N}{}^{(1)}_{(r)j}}{\partial t} - H_{11}^1 \underset{1}{T}{}^{(1)}_{(r)j}, \quad \underset{1}{R}{}^{(1)}_{(r)ij} = -\mathfrak{R}_{rij}^k p_k^1 + \left[\underset{1}{T}{}^{(1)}_{(r)i|j} - \underset{1}{T}{}^{(1)}_{(r)j|i}\right],$$

where

$$\mathfrak{R}_{kij}^r = \frac{\partial \gamma_{ki}^r}{\partial x^j} - \frac{\partial \gamma_{kj}^r}{\partial x^i} + \gamma_{ki}^p \gamma_{pj}^r - \gamma_{kj}^p \gamma_{pi}^r, \quad \underset{1}{T}{}^{(1)}_{(i)j} = h^{11}\left(X_{i1\cdot j} + X_{j1\cdot i}\right).$$

Moreover, all the curvature d-tensors of the Cartan canonical connection $C\Gamma(N)$ of the least squares Hamiltonian (5.3) are zero, except

$$\mathbf{R}_{(i)(1)jk}^{(1)(l)} = -R_{(i)(1)jk}^{(1)(l)} = -R_{ijk}^l := -\Re_{ijk}^l.$$

Remark 5.2 If we have $h_{11} = 1$ and $\varphi_{ij} = \delta_{ij}$, we find the torsion components produced by the least squares Hamiltonian function (5.3) as being the following:

$$\mathbf{R}_{(r)1j}^{(1)} = -\left(\frac{\partial^2 X_{(1)}^{(r)}}{\partial t \partial x^j} + \frac{\partial^2 X_{(1)}^{(j)}}{\partial t \partial x^r} \right), \quad \mathbf{R}_{(r)ij}^{(1)} = \frac{\partial^2 X_{(1)}^{(i)}}{\partial x^r \partial x^j} - \frac{\partial^2 X_{(1)}^{(j)}}{\partial x^r \partial x^i}.$$

Moreover, all the curvature d-tensors produced by the least squares Hamiltonian (5.3) are zero.

The components $F_{(1)j}^{(i)}$ and $f_{(1)(1)}^{(i)(j)}$ of the momentum electromagnetic-like field $\mathbb{F}$, attached to the least squares Hamiltonian function (5.3), are expressed by

$$F_{(1)j}^{(i)} = \frac{1}{8} \left[\varphi^{jk} X_{k1\cdot i} - \varphi^{ik} X_{k1\cdot j} + \varphi^{jk} X_{i1\cdot k} - \varphi^{ik} X_{j1\cdot k} \right], \quad f_{(a)(b)}^{(i)(j)} = 0.$$

Remark 5.3 If we have $h_{11} = 1$ and $\varphi_{ij} = \delta_{ij}$, we find that $F_{(1)j}^{(i)} = 0$, that is, the momentum electromagnetic-like field in this case is trivial, i.e., $\mathbb{F} = 0$.

5.2   Hamilton Geometrization of an Economy Dynamical System

We study now the dynamical of competition between two economical sectors governed by the first-order differential system (see Ferrara-Niglia [29], Udriște-Ferrara-Opriş [30], and Neagu [31])

$$\begin{cases} \dfrac{dE_1}{dt} = g_1 E_1 \left(1 - \dfrac{E_1}{K_1} - \beta_1 \dfrac{E_2}{K_1} \right) \\[2mm] \dfrac{dE_2}{dt} = g_2 E_2 \left(1 - \dfrac{E_2}{K_2} - \beta_2 \dfrac{E_1}{K_2} \right), \end{cases} \tag{5.6}$$

where

- E_1 and E_2 are two populations of new firms born in the above economical sectors;
- g_1 and g_2 are strictly positive constants representing the growth rates of the two economical sectors;
- K_1 and K_2 are strictly positive constants representing the investments of capitals;
- β_1 and β_2 are strictly positive constants representing the competitive interaction coefficients.

The differential system (5.6) can be regarded on the 1-jet space $J^1(\mathbb{R}, M)$, whose coordinates are $(t, x^1 = E_1, x^2 = E_2, y_1^1 = dE_1/dt, y_1^2 = dE_2/dt)$.

In this context, the solutions of class C^2 of system (5.6) are the global minimum points of the least square Lagrangian (see [31])

$$
L = \left(y_1^1 - X_{(1)}^{(1)}(t, E_1, E_2) \right)^2 + \left(y_1^2 - X_{(1)}^{(2)}(t, E_1, E_2) \right)^2
$$

$$
= \sum_{i,j=1}^{2} 1 \cdot \delta_{ij} \left(y_1^i - X_{(1)}^{(i)}(t, E_1, E_2) \right) \left(y_1^j - X_{(1)}^{(j)}(t, E_1, E_2) \right),
$$

where

$$
\begin{cases}
X_{(1)}^{(1)}(t, E_1, E_2) = g_1 E_1 \left(1 - \dfrac{E_1}{K_1} - \beta_1 \dfrac{E_2}{K_1} \right) \\[3mm]
X_{(1)}^{(2)}(t, E_1, E_2) = g_2 E_2 \left(1 - \dfrac{E_2}{K_2} - \beta_2 \dfrac{E_1}{K_2} \right),
\end{cases}
$$

whose corresponding least squares Hamiltonian is given by

$$
H = \frac{\delta^{ij}}{4} p_i^1 p_j^1 + X_{(1)}^{(k)} p_k^1 = \frac{1}{4} \left[\left(p_1^1 \right)^2 + \left(p_2^1 \right)^2 \right] + X_{(1)}^{(1)} p_1^1 + X_{(1)}^{(2)} p_2^1.
$$

Applying the preceding geometrical theory, it follows that, using the Jacobian notation

$$
J(X) = \left(\frac{\partial X_{(1)}^{(i)}}{\partial E_j} \right)_{i,j=\overline{1,2}} = \begin{pmatrix} g_1 - 2g_1 \dfrac{E_1}{K_1} - g_1\beta_1 \dfrac{E_2}{K_1} & -g_1\beta_1 \dfrac{E_1}{K_1} \\[3mm] -g_2\beta_2 \dfrac{E_2}{K_2} & g_2 - 2g_2 \dfrac{E_2}{K_2} - g_2\beta_2 \dfrac{E_1}{K_2} \end{pmatrix},
$$

we find the following geometrical objects associated with the dynamical system (5.6) (here we have $i, j \in \{1, 2\}$):

1. The coefficients of the *canonical nonlinear connection* produced by the dynamical system (5.6) are given by the temporal components $\underset{1}{N}_{(i)1}^{(1)} = 0$, and the spatial components are the entries of the symmetric matrix

$$
\underset{2}{N} = \left(\underset{2}{N}_{(i)j}^{(1)} \right) = \left(\frac{\partial X_{(1)}^{(i)}}{\partial E_j} + \frac{\partial X_{(1)}^{(j)}}{\partial E_i} \right) = J(X) + J(X)^T
$$

$$
= \begin{pmatrix} 2\left(g_1 - 2g_1 \dfrac{E_1}{K_1} - g_1\beta_1 \dfrac{E_2}{K_1} \right) & -g_1\beta_1 \dfrac{E_1}{K_1} - g_2\beta_2 \dfrac{E_2}{K_2} \\[3mm] -g_1\beta_1 \dfrac{E_1}{K_1} - g_2\beta_2 \dfrac{E_2}{K_2} & 2\left(g_2 - 2g_2 \dfrac{E_2}{K_2} - g_2\beta_2 \dfrac{E_1}{K_2} \right) \end{pmatrix}.
$$

Moreover, all coefficients of the *Cartan canonical connection* $C\Gamma(N)$ of the least squares Hamiltonian function (5.3) are zero.

2. The nonzero *torsion components* produced by the dynamical system (5.6) are the entries of the matrices:

$$
\mathbf{R}_{(1)} = \left(\mathbf{R}^{(1)}_{(1)ij} \right) = \left(\frac{\partial^2 X^{(i)}_{(1)}}{\partial E_1 \partial E_j} - \frac{\partial^2 X^{(j)}_{(1)}}{\partial E_1 \partial E_i} \right)
$$

$$
= \frac{d}{dE_1} \left[J(X) - J(X)^T \right] = \begin{pmatrix} 0 & -\dfrac{g_1 \beta_1}{K_1} \\ \dfrac{g_1 \beta_1}{K_1} & 0 \end{pmatrix};
$$

$$
\mathbf{R}_{(2)} = \left(\mathbf{R}^{(1)}_{(2)ij} \right) = \left(\frac{\partial^2 X^{(i)}_{(1)}}{\partial E_2 \partial E_j} - \frac{\partial^2 X^{(j)}_{(1)}}{\partial E_2 \partial E_i} \right)
$$

$$
= \frac{d}{dE_2} \left[J(X) - J(X)^T \right] = \begin{pmatrix} 0 & \dfrac{g_2 \beta_2}{K_2} \\ -\dfrac{g_2 \beta_2}{K_2} & 0 \end{pmatrix}.
$$

Moreover, all the *curvature d-tensors* produced by the dynamical system (5.6) are zero.

Time-Dependent Hamiltonian of Electrodynamics 6

Abstract

In Chap. 6, we develop the distinguished Riemannian differential geometry (in the sense of d-connections, d-torsions, d-curvatures, and the geometrical Maxwell-like and Einstein-like equations) for a time-dependent Hamiltonian of momenta which governs the electrodynamics phenomena.

6.1   Introduction

The geometrical development from this chapter follows Balan-Neagu-Oană-Ovsiyuk's paper [32]. An extension of classical mechanics for a non-relativistic particle with a fixed mass m in the presence of the external non-autonomous electromagnetic field $A_i(t, x^i)$ is physically studied by Landau and Lifshitz in [33, 34]. In the same direction, in the classical Lagrange geometry developed on the tangent bundle TM, the Lagrangian $L : TM \rightarrow \mathbb{R}$ that governs the electrodynamics phenomena is given by (see Miron [35])

$$L(x, y) = mc\varphi_{ij}(x)y^i y^j + \frac{2e}{m}A_i(x)y^i + U(x),$$

where $\varphi_{ij}(x)$ is a semi-Riemannian metric tensor on M representing the *gravitational potentials*; $A_i(x)$ is a covector field on M describing *electromagnetic potentials*; $U(x)$ is a function; and $m \neq 0$, c, and e are constants of the physics as the *mass, speed of light*, or *electric charge*. In this way, we recall that a jet extension of the Lagrangian function of electrodynamics $L : J^1(\mathbb{R}, M) \rightarrow \mathbb{R}$ is set by (see Neagu [36])

$$L(t, x^k, y_1^k) = mch^{11}(t)\varphi_{ij}(x)y_1^i y_1^j + \frac{2e}{m}A_{(i)}^{(1)}(t, x)y_1^i + \mathsf{P}(t, x), \tag{6.1}$$

© The Author(s), under exclusive license to Springer Nature Switzerland AG 2022

M. Neagu and A. Oană, *Dual Jet Geometrization for Time-Dependent Hamiltonians and Applications*, Synthesis Lectures on Mathematics & Statistics,
https://doi.org/10.1007/978-3-031-08885-8_6

where $h_{11}(t)$ (respectively, $\varphi_{ij}(x)$) is a semi-Riemannian metric on the time manifold $\mathbb{R}$ (respectively, spatial manifold M), $A_{(i)}^{(1)}(t, x)$ is a distinguished tensor on $J^1(\mathbb{R}, M)$, and $P(t, x)$ is a smooth function on the product manifold $\mathbb{R} \times M$.

Via the Legendre transformations

$$p_i^1 = \frac{\partial L}{\partial y_1^i} = 2mch^{11}\varphi_{ij}y_1^j + \frac{2e}{m}A_{(i)}^{(1)} \Leftrightarrow$$

$$\Leftrightarrow y_1^i = \frac{h_{11}\varphi^{ij}}{2mc}\left[p_j^1 - \frac{2e}{m}A_{(j)}^{(1)}\right],$$

the jet time-dependent Lagrangian function of electrodynamics (6.1) leads us to the Hamiltonian function of momenta

$$H = p_i^1 y_1^i - L = \frac{1}{4mc}h_{11}\varphi^{ij}p_i^1 p_j^1 - \frac{e}{m^2 c}h_{11}\varphi^{ij}A_{(j)}^{(1)}p_i^1 + \frac{e^2}{m^3 c}\|A\|^2 - \mathrm{P}, \qquad (6.2)$$

where $H : J^{1*}(\mathbb{R}, M) \to \mathbb{R}$, and $\|A\|^2(t, x) = h_{11}\varphi^{ij}A_{(i)}^{(1)}A_{(j)}^{(1)}$. The pair

$$\mathcal{EDH}^n = (J^{1*}(\mathbb{R}, M), H),$$

where H is given by (6.2), is called the *autonomous time-dependent Hamilton space of electrodynamics*. Now, using as a pattern Miron's geometrical ideas from the works [37] on TM and [38, 39] on T^*M, which were extended on 1-jet spaces by Neagu [36] and their duals by Atanasiu et al. [40], the distinguished Riemannian geometry for the particular momentum Hamiltonian function (6.2) (which governs the *time-dependent momentum electrodynamics*) can be constructed on the dual 1-jet space $J^{1*}(\mathbb{R}, M)$ (see Part I of this monograph).

6.2   The Time-Dependent Hamilton Space of Electrodynamics

To start our Hamiltonian geometrical development for time-dependent electrodynamics, let us consider on the dual 1-jet space $E^* = J^{1*}(\mathbb{R}, M)$ the *vertical fundamental metrical d-tensor*

$$G_{(1)(1)}^{(i)(j)} = \frac{1}{2}\frac{\partial^2 H}{\partial p_i^1 \partial p_j^1} = \tilde{h}_{11}(t)\varphi^{ij}(x^k),$$

where $\tilde{h}_{11}(t) := (4mc)^{-1} \cdot h_{11}(t)$. Let $H_{11}^1(t) = (h^{11}/2)(dh_{11}/dt)$ (respectively, $\gamma_{ij}^k(x)$) be the Christoffel symbols of the metric $h_{11}(t)$ (respectively, $\varphi_{ij}(x)$). Obviously, if $\widetilde{H}_{11}^1$ is the Christoffel symbol of the semi-Riemannian metric $\tilde{h}_{11}(t)$, then we have $\widetilde{H}_{11}^1 = H_{11}^1$. In this context, by direct computations, we find (see general Formulas (4.4))

Theorem 6.1 *The pair of local functions*

$$N = \left(\underset{1}{N}{}^{(1)}_{(i)1},\ \underset{2}{N}{}^{(1)}_{(i)j} \right)$$

on the dual 1-jet space E^*, *which are given by*

$$\underset{1}{N}{}^{(1)}_{(i)1} = H^1_{11} p^1_i,\quad \underset{2}{N}{}^{(1)}_{(i)j} = \gamma^r_{ij}\left[\frac{2e}{m} A^{(1)}_{(r)} - p^1_r \right] - \frac{e}{m}\left[\frac{\partial A^{(1)}_{(i)}}{\partial x^j} + \frac{\partial A^{(1)}_{(j)}}{\partial x^i} \right], \tag{6.3}$$

represents a nonlinear connection on E^*. *This nonlinear connection is called the* **canonical nonlinear connection of the time-dependent Hamilton space of electrodynamics** $\mathcal{E}DH^n$.

Now, let $\{\delta/\delta t,\ \delta/\delta x^i,\ \partial/\partial p^1_i\} \subset \mathcal{X}(E^*)$ and $\{dt, dx^i, \delta p^1_i\} \subset \mathcal{X}^*(E^*)$ be the adapted bases produced by the nonlinear connection (6.3), where

$$\frac{\delta}{\delta t} = \frac{\partial}{\partial t} - \underset{1}{N}{}^{(1)}_{(r)1}\frac{\partial}{\partial p^1_r},\quad \frac{\delta}{\delta x^i} = \frac{\partial}{\partial x^i} - \underset{2}{N}{}^{(1)}_{(r)i}\frac{\partial}{\partial p^1_r},$$

$$\delta p^1_i = dp^1_i + \underset{1}{N}{}^{(1)}_{(i)1}dt + \underset{2}{N}{}^{(1)}_{(i)r}dx^r. \tag{6.4}$$

Using the adapted bases (6.4), by direct local computations, we can determine the adapted components of the Cartan canonical connection of the space $\mathcal{E}DH^n$, together with its local d-torsions and d-curvatures (see the general formulas from Tables 3.1 and 3.2).

Theorem 6.2 (i) *The canonical Cartan connection of the autonomous time-dependent Hamilton space of electrodynamics* $\mathcal{E}DH^n$ *is defined by adapted components*

$$C\Gamma(N) = \left(H^1_{11} = H^1_{11},\ A^i_{j1} = 0,\ H^i_{jk} = \gamma^i_{jk},\ C^{i(k)}_{j(1)} = 0 \right).$$

(ii) *The torsion* **T** *of the canonical Cartan connection of the space* $\mathcal{E}DH^n$ *is determined by* **two** *effective adapted components:*

$$R^{(1)}_{(r)1j} = -\frac{2e}{m}\gamma^s_{rj}A^{(1)}_{(s);1} + \frac{e}{m}\left[\frac{\partial A^{(1)}_{(r)}}{\partial x^j} + \frac{\partial A^{(1)}_{(j)}}{\partial x^r} \right]_{;1},$$

$$R^{(1)}_{(r)ij} = \mathfrak{R}^s_{rij}\left[\frac{2e}{m}A^{(1)}_{(s)} - p^1_s \right] - \frac{e}{m}\left[\frac{\partial A^{(1)}_{(i)}}{\partial x^j} - \frac{\partial A^{(1)}_{(j)}}{\partial x^i} \right]_{:r}, \tag{6.5}$$

where $\mathfrak{R}^k_{rij}(x)$ *are the local curvature tensors of the semi-Riemannian metric* $\varphi_{ij}(x)$, *and* "$_{;1}$" *and* "$_{:k}$" *represent the following* **generalized Levi-Civita covariant derivatives:**

- *the $\mathbb{R}$-generalized Levi-Civita covariant derivative:*

$$T^{1i(1)(r)\dots}_{1j(l)(1)\dots;1} \overset{def}{=} \frac{\partial T^{1i(1)(r)\dots}_{1j(l)(1)\dots}}{\partial t} + T^{1i(1)(r)\dots}_{1j(l)(1)\dots} H^1_{11} + T^{1i(1)(r)\dots}_{1j(l)(1)\dots} H^1_{11} + \dots$$
$$\dots - T^{1i(1)(r)\dots}_{1j(l)(1)\dots} H^1_{11} - T^{1i(1)(r)\dots}_{1j(l)(1)\dots} H^1_{11} - \dots$$

- *the M-generalized Levi-Civita covariant derivative:*

$$T^{1i(1)(r)\dots}_{1j(l)(1)\dots;k} \overset{def}{=} \frac{\partial T^{1i(1)(r)\dots}_{1j(l)(1)\dots}}{\partial x^k} + T^{1s(1)(r)\dots}_{1j(l)(1)\dots} \gamma^i_{sk} + T^{1i(1)(s)\dots}_{1j(l)(1)\dots} \gamma^r_{sk} + \dots$$
$$\dots - T^{1i(1)(r)\dots}_{1s(l)(1)\dots} \gamma^s_{jk} - T^{1i(1)(r)\dots}_{1j(s)(1)\dots} \gamma^s_{lk} - \dots .$$

*(iii) The curvature $\mathbf{R}$ of the Cartan connection of the space $\mathcal{EDH}^n$ is given by **one** effective adapted component: $R^l_{ijk} = \mathfrak{R}^l_{ijk}$.*

6.3   Momentum Electromagnetic-Like Geometrical Model

To expose our geometrical electromagnetic-like theory on the time-dependent Hamilton space of electrodynamics $\mathcal{EDH}^n$, we emphasize that, by simple direct calculations, we obtain

Proposition 6.1 *The **metrical deflection d-tensors** of the space $\mathcal{EDH}^n$ are given by the following formulas:*

$$\Delta^{(i)}_{(1)j} = [\tilde{h}_{11}\varphi^{ir} p^1_r]_{|j} = \frac{e}{4m^2c} h_{11}\varphi^{ir} \left[A^{(1)}_{(r):j} + A^{(1)}_{(j):r} \right],$$

$$\Delta^{(i)}_{(1)1} = [\tilde{h}_{11}\varphi^{ir} p^1_r]_{/1} = 0, \tag{6.6}$$

$$\vartheta^{(i)(j)}_{(1)(1)} = [\tilde{h}_{11}\varphi^{ir} p^1_r]|^{(j)}_{(1)} = \frac{1}{4mc} h_{11}\varphi^{ij},$$

where "$_{/1}$", "$_{|j}$", and "$|^{(1)}_{(j)}$" are the local covariant derivatives induced by the Cartan canonical connection $C\Gamma(N)$.

Moreover, following some general formulas from Part I, we introduce the following abstract geometric-physical concept.

Definition 6.1 The distinguished 2-form on the 1-jet space $E^* = J^{1*}(\mathbb{R}, M)$ locally defined by

$$\mathbb{F} = F^{(i)}_{(1)j}\delta p^1_i \wedge dx^j + f^{(i)(j)}_{(1)(1)}\delta p^1_i \wedge \delta p^1_j,$$

where

$$F^{(i)}_{(1)j} = \frac{1}{2}\left[\Delta^{(i)}_{(1)j} - \Delta^{(j)}_{(1)i}\right] = \frac{e}{8m^2c} \cdot \mathcal{A}_{\{i,j\}}\left\{h_{11}\varphi^{ir}\left[A^{(1)}_{(r):j} + A^{(1)}_{(j):r}\right]\right\},$$

$$f^{(i)(j)}_{(1)(1)} = \frac{1}{2}\left[\vartheta^{(i)(j)}_{(1)(1)} - \vartheta^{(j)(i)}_{(1)(1)}\right] = 0$$

(6.7)

is called the **distinguished momentum electromagnetic field associated with the autonomous time-dependent Hamilton space of electrodynamics $\mathcal{EDH}^n$**.

By particularizing on the space $\mathcal{EDH}^n$ the geometrical Maxwell-like equations of the momentum electromagnetic field that governs a general time-dependent Hamilton space H^n, we get

Theorem 6.3 *The momentum electromagnetic components (6.7) of the autonomous time-dependent Hamilton space of electrodynamics $\mathcal{EDH}^n$ are governed by the following **geometrical Maxwell-like equations**:*

$$\begin{cases} F^{(i)}_{(1)j/1} = F^{(i)}_{(1)j;1} = \dfrac{e \cdot h_{11}}{8m^2c} \cdot \mathcal{A}_{\{i,j\}}\varphi^{ir}\left\{\left[\dfrac{\partial A^{(1)}_{(r)}}{\partial x^j} + \dfrac{\partial A^{(1)}_{(j)}}{\partial x^r}\right]_{;1} - 2\gamma^s_{rj}A^{(1)}_{(s);1}\right\}, \\[4mm] \displaystyle\sum_{\{i,j,k\}} F^{(i)}_{(1)j|k} = \sum_{\{i,j,k\}} F^{(i)}_{(1)j;k} = -\dfrac{h_{11}}{8mc} \cdot \sum_{\{i,j,k\}}\left\{\left[\varphi^{sr}\mathfrak{R}^i_{rjk} - \varphi^{ir}\mathfrak{R}^s_{rjk}\right]p^1_s + \right. \\[4mm] \qquad \left. + \dfrac{e}{m}\varphi^{ir}\left[2\mathfrak{R}^s_{rjk}A^{(1)}_{(s)} - \left(\dfrac{\partial A^{(1)}_{(j)}}{\partial x^k} - \dfrac{\partial A^{(1)}_{(k)}}{\partial x^j}\right)_{:r}\right]\right\}, \end{cases}$$

where $\mathcal{A}_{\{i,j\}}$ represents an alternate sum and $\sum_{\{i,j,k\}}$ represents a cyclic sum.

6.4   Momentum Gravitational-Like Geometrical Model

To describe our geometrical Hamiltonian momentum gravitational theory on the autonomous time-dependent Hamilton space of electrodynamics $\mathcal{EDH}^n$, we recall that the metrical d-tensor $G^{(i)(j)}_{(1)(1)} = \widetilde{h}_{11}(t)\varphi^{ij}(x)$ and the canonical nonlinear connection (6.3) produce a momentum gravitational $\widetilde{h}$-potential $\mathbb{G}$ on the 1-jet space E^*, locally defined by

$$\mathbb{G} = \widetilde{h}_{11}dt \otimes dt + \varphi_{ij}dx^i \otimes dx^j + \widetilde{h}_{11}\varphi^{ij}\delta p^1_i \otimes \delta p^1_j.$$

To analyze the corresponding local geometrical Einstein-like equations (together with their momentum conservation laws) in the adapted basis $\{X_A\} = \{\delta/\delta t,\ \delta/\delta x^i,\ \partial/\partial p^1_i\}$, let

$C\Gamma(N) = (H^1_{11}, 0, \gamma^i_{jk}, 0)$ be the Cartan canonical connection of the space $\mathcal{E}\mathcal{D}H^n$. Taking into account the expressions of its adapted curvature d-tensors on the space $\mathcal{E}\mathcal{D}H^n$, we find

Theorem 6.4 *The Ricci tensor* $\mathrm{Ric}(C\Gamma(N))$ *of the space* $\mathcal{E}\mathcal{D}H^n$ *is characterized only by one effective local adapted Ricci d-tensor:* $\mathfrak{R}_{ij} = \mathfrak{R}^r_{ijr}$.

The scalar curvature $Sc(C\Gamma(N))$ of the Cartan connection of the space $\mathcal{E}\mathcal{D}H^n$ is given by $Sc(C\Gamma(N)) = \mathfrak{R}$, where $\mathfrak{R} = \varphi^{ij}\mathfrak{R}_{ij}$ is the scalar curvature of the semi-Riemannian metric $\varphi_{ij}(x)$. By particularizing on the space $\mathcal{E}\mathcal{D}H^n$ the geometrical Einstein-like equations and the momentum conservation laws that govern an arbitrary time-dependent Hamilton space H^n, we get

Theorem 6.5 *The local **geometrical Einstein-like equations**, that govern the momentum gravitational potential of the space* $\mathcal{E}\mathcal{D}H^n$, *have the following form:*

$$
\begin{cases}
\mathfrak{R}_{ij} - \dfrac{\mathfrak{R}}{2}\varphi_{ij} = \mathcal{K}\mathbb{T}_{ij}, \\[2mm]
0 = \mathbb{T}_{1i}, \quad 0 = \mathbb{T}_{i1}, \quad 0 = \mathbb{T}^{(i)}_{(1)1}, \quad -\mathfrak{R}h_{11} = 8mc \cdot \mathcal{K}\mathbb{T}_{11}, \\[2mm]
0 = \mathbb{T}^{(j)}_{1(1)}, \; 0 = \mathbb{T}^{(j)}_{i(1)}, \; 0 = \mathbb{T}^{(i)}_{(1)j}, \quad -\mathfrak{R}h_{11}\varphi^{ij} = 8mc \cdot \mathcal{K}\mathbb{T}^{(i)(j)}_{(1)(1)},
\end{cases}
\tag{6.8}
$$

where $\mathbb{T}_{AB}$, $A, B \in \left\{1, i, \begin{smallmatrix}(i)\\(1)\end{smallmatrix}\right\}$, *are the adapted components of the momentum stress-energy d-tensor of matter* $\mathbb{T}$, *and* $\mathcal{K}$ *is the Einstein constant.*

As a consequence, setting $\mathfrak{R}^r_j = \varphi^{rs}\mathfrak{R}_{sj}$, then the *momentum conservation laws* of the geometrical Einstein-like equations (6.8) take the following form:

$$
\left[\mathfrak{R}^r_j - \dfrac{\mathfrak{R}}{2}\delta^r_j\right]_{|r} = 0.
$$

! Open problem

From physical point of view, an open problem is to describe the properties of such mechanical models which correspond to the introduced above momenta-depending geometrical objects.

The Geometry of Conformal Hamiltonian of the Time-Dependent Coupled Harmonic Oscillators

7

Abstract

In this chapter, we construct the d-geometry (in the sense of d-connections, d-torsions, d-curvatures, momentum geometrical gravitational-like, and electromagnetic-like theories) for the conformal Hamiltonian of the time-dependent coupled oscillators on the dual 1-jet space $J^{1*}(\mathbb{R}, \mathbb{R}^2)$.

7.1 Introduction

The geometrical ideas from this chapter follow the paper Raeisi-Dehkordi and Neagu [41]. The model of time-dependent coupled oscillators is used (see Macedo and Guedes [42]) to investigate the dynamics of charged particle motion in the presence of time-varying magnetic fields. At the same time, the model of coupled harmonic oscillator has also been widely used to study the quantum effects in mesoscopic coupled electric circuits. For more details, see also [42].

If m_i ($i = \overline{1,2}$), ω_i ($i = \overline{1,2}$), and $k(t)$ are the time-dependent mass, frequency, and the coupling parameter, respectively, then the conformal Hamiltonian of the time-dependent coupled harmonic oscillators is given on the dual 1-jet space $J^{1*}(\mathbb{R}, \mathbb{R}^2)$ by [42]

$$
\begin{aligned}
H(t, x, p) &= h_{11}(t)\, e^{\sigma(x)} \left[\frac{\left(p_1^1\right)^2}{m_1(t)} + \frac{\left(p_2^1\right)^2}{m_2(t)} \right] + \mathcal{F}(t, x) \\
&= h_{11}(t)\, e^{\sigma(x)} \frac{\delta_{ij}}{m_i(t)} p_i^1 p_j^1 + \mathcal{F}(t, x) \\
&= h_{11}(t)\, g^{ij}(t, x) p_i^1 p_j^1 + \mathcal{F}(t, x),
\end{aligned}
\tag{7.1}
$$

where $\sigma : \mathbb{R}^2 \to \mathbb{R}$ is a smooth conformal function on $\mathbb{R}^2$, h_{11} is a Riemannian metric on $\mathbb{R}$, $(t, x, p) = \left(t, x_1, x_2, p_1^1, p_2^1\right)$ are the coordinates of the space $J^{1*}(\mathbb{R}, \mathbb{R}^2)$, and

© The Author(s), under exclusive license to Springer Nature Switzerland AG 2022
M. Neagu and A. Oană, *Dual Jet Geometrization for Time-Dependent Hamiltonians and Applications*, Synthesis Lectures on Mathematics & Statistics,
https://doi.org/10.1007/978-3-031-08885-8_7

$$\mathcal{F}(t, x) = \frac{m_1(t)\,\omega_1^2(t)\,x_1^2}{2} + \frac{m_2(t)\,\omega_2^2(t)\,x_2^2}{2} + \frac{k(t)\,(x_2 - x_1)^2}{2}.$$

We recall that the dual jet coordinates (t, x, p) transform by the following rules:

$$t = t(t), \quad \tilde{x}^i = \tilde{x}^i\left(x^j\right), \quad \tilde{p}_i^1 = \frac{\partial x^j}{\partial \tilde{x}^i}\frac{d\tilde{t}}{dt}p_j^1, \tag{7.2}$$

where $i, j = \overline{1, 2}$, $\dfrac{d\tilde{t}}{dt} \neq 0$, and $rank\left(\dfrac{\partial \tilde{x}^i}{\partial x^j}\right)_{i, j = \overline{1, 2}} = 2$.

7.2   The Canonical Nonlinear Connection

The *fundamental metrical d-tensor* induced by the conformal Hamiltonian metric (7.1) is defined by

$$g^{ij}(t, x) = \frac{h^{11}}{2}\frac{\partial^2 H}{\partial p_i^1 \partial p_j^1} = \frac{\delta_{ij}}{m_i(t)}e^{\sigma(x)} \Rightarrow g_{jk}(t, x) = \delta_{jk}m_k(t)e^{-\sigma(x)}. \tag{7.3}$$

If we use the notations

$$K_{11}^1 = \frac{h^{11}}{2}\frac{dh_{11}}{dt},$$

$$\Gamma_{ij}^k = \frac{g^{kl}}{2}\left(\frac{\partial g_{li}}{\partial x^j} + \frac{\partial g_{lj}}{\partial x^i} - \frac{\partial g_{ij}}{\partial x^l}\right) = \frac{1}{2}\left(-\delta_i^k\sigma_j - \delta_j^k\sigma_i + \delta_{ij}\frac{m_j(t)}{m_k(t)}\sigma_k\right), \tag{7.4}$$

where $h^{11} = \dfrac{1}{h_{11}} > 0$ and $\sigma_i = \dfrac{\partial \sigma}{\partial x^i}$, then we have the following geometrical result.

Proposition 7.1 *For the conformal Hamiltonian metric (7.1), the canonical nonlinear connection on the dual 1-jet space $J^{1*}(\mathbb{R}, \mathbb{R}^2)$ has the following components:*

$$N = \left(\underset{1}{N}^{(1)}_{(i)1} = K_{11}^1 p_i^1, \ \underset{2}{N}^{(1)}_{(i)j} = -\Gamma_{ij}^k p_k^1\right). \tag{7.5}$$

In other words, we have

$$\underset{2}{N}^{(1)}_{(i)j} = \frac{1}{2}\left(\sigma_i p_j^1 + \sigma_j p_i^1 - \delta_{ij}\frac{m_j(t)}{m_k(t)}\sigma_k p_k^1\right).$$

Proof The canonical nonlinear connection produced by H on the dual 1-jet space $J^{1*}(\mathbb{R}, \mathbb{R}^2)$ has the following components: $N^{(1)}_{1\,(i)1} = K^1_{11} p^1_i$ and

$$
N^{(1)}_{2\,(i)j} = \frac{h^{11}}{4} \left[\frac{\partial g_{ij}}{\partial x^k} \frac{\partial H}{\partial p^1_k} - \frac{\partial g_{ij}}{\partial p^1_k} \frac{\partial H}{\partial x^k} + g_{ik} \frac{\partial^2 H}{\partial x^j \partial p^1_k} + g_{jk} \frac{\partial^2 H}{\partial x^i \partial p^1_k} \right].
$$

So, by direct computations, we obtain (7.5). $\qquad\square$

7.3 Cartan Canonical Connection. d-Torsions and d-Curvatures

The nonlinear connection N produces the following *dual adapted bases* of d-vector fields and d-covector fields:

$$
\left\{ \frac{\delta}{\delta t}, \frac{\delta}{\delta x^i}, \frac{\partial}{\partial p^1_r} \right\} \subset X\left(J^{1*}(\mathbb{R}, \mathbb{R}^2)\right), \quad \left\{ dt, dx^i, \delta p^1_i \right\} \subset X^*\left(J^{1*}(\mathbb{R}, \mathbb{R}^2)\right), \tag{7.6}
$$

where

$$
\frac{\delta}{\delta t} = \frac{\partial}{\partial t} - N^{(1)}_{1\,(r)1} \frac{\partial}{\partial p^1_r} = \frac{\partial}{\partial t} - K^1_{11} p^1_r \frac{\partial}{\partial p^1_r}, \tag{7.7}
$$

$$
\frac{\delta}{\delta x^i} = \frac{\partial}{\partial x^i} - N^{(1)}_{2\,(r)i} \frac{\partial}{\partial p^1_r} = \frac{\partial}{\partial x^i} + \Gamma^s_{ri} p^1_s \frac{\partial}{\partial p^1_r}, \tag{7.8}
$$

$$
\delta p^1_i = dp^1_i + N^{(1)}_{1\,(i)1} dt + N^{(1)}_{2\,(i)r} dx^r.
$$

We recall that the naturalness of the above geometrical adapted bases is coming from the fact that, via a transformation of coordinates (7.2), their elements transform as classical tensors. Therefore, the description of all subsequent geometrical objects on the dual 1-jet space $J^{1*}(\mathbb{R}, \mathbb{R}^2)$ (e.g., the Cartan canonical N-linear connection, its torsion and curvature, etc.) will be done in local adapted components. As a result, by direct computations, we obtain the following geometrical result.

Proposition 7.2 *The Cartan canonical N-linear connection produced by the conformal Hamiltonian metric (7.1) has the following adapted local components:*

$$
C\Gamma(N) = \left(K^1_{11}, \ A^i_{j1} = \delta^i_j \frac{m'_j(t)}{2m_i(t)}, \ H^i_{jk} = \Gamma^i_{jk}, \ C^{i(k)}_{j(1)} = 0 \right). \tag{7.9}
$$

Proof The adapted components of Cartan canonical connection are given by the following formulas:

$$A^i_{j1} = \frac{g^{il}}{2} \frac{\delta g_{lj}}{\delta t} = \frac{g^{il}}{2} \frac{dg_{lj}}{dt} = \frac{\delta^i_j}{2m_i(t)} \frac{dm_j}{dt} = \delta^i_j \frac{m'_j(t)}{2m_i(t)}$$

$$H^i_{jk} = \frac{g^{ir}}{2} \left(\frac{\delta g_{jr}}{\delta x^k} + \frac{\delta g_{kr}}{\delta x^j} - \frac{\delta g_{jk}}{\delta x^r} \right) = \Gamma^i_{jk},$$

$$C^{j(k)}_{i(1)} = -\frac{g_{ir}}{2} \left(\frac{\partial g^{jr}}{\partial p^1_k} + \frac{\partial g^{kr}}{\partial p^1_j} - \frac{\partial g^{jk}}{\partial p^1_r} \right) = 0.$$

Using the derivative operators (7.7) and (7.8), the direct calculations lead us to the desired results.   □

Proposition 7.3 *The Cartan canonical connection of the conformal Hamiltonian metric (7.1) has four effective d-torsions:*

$$T^r_{1j} = -A^r_{j1}, \quad P^{(1)\,(j)}_{(r)1(1)} = A^j_{r1}, \quad R^{(1)}_{(r)1j} = \frac{\partial \Gamma^s_{rj}}{\partial t} p^1_s, \quad R^{(1)}_{(r)ij} = -\Re^{\,k}_{rij} p^1_k,$$

where

$$\Re^l_{ijk} = \frac{\partial \Gamma^l_{ij}}{\partial x^k} - \frac{\partial \Gamma^l_{ik}}{\partial x^j} + \Gamma^s_{ij} \Gamma^l_{sk} - \Gamma^s_{ik} \Gamma^l_{sj}.$$

Proof The Cartan canonical connection on the dual 1-jet space $J^{1*}(\mathbb{R}, \mathbb{R}^2)$ generally has six effective local d-tensors of torsion. For our particular Cartan canonical connection (7.9), these reduce only to four (the other two are zero):

$$T^r_{1j} = -A^r_{j1}, \quad P^{r(j)}_{i(1)} = C^{r(j)}_{i(1)} = 0,$$

$$P^{(1)\,(j)}_{(r)1(1)} = \frac{\partial \overset{(1)}{N}_{(r)1}}{\partial p^1_j} + A^j_{r1} - \delta^j_r K^1_{11} = A^j_{r1}, \quad P^{(1)\,(j)}_{(r)i(1)} = \frac{\partial \overset{(1)}{N}_{(r)i}}{\partial p^1_j} + \Gamma^j_{ri} = 0,$$

$$R^{(1)}_{(r)1j} = \frac{\delta \overset{(1)}{N}_{(r)1}}{\delta x^j} - \frac{\delta \overset{(1)}{N}_{(r)j}}{\delta t} = -\frac{\partial \overset{(1)}{N}_{(r)j}}{\partial t} = \frac{\partial \Gamma^s_{rj}}{\partial t} p^1_s,$$

$$R^{(1)}_{(r)ij} = \frac{\delta \overset{(1)}{N}_{(r)i}}{\delta x^j} - \frac{\delta \overset{(1)}{N}_{(r)j}}{\delta x^i} = -\Re^{\,k}_{rij} p^1_k. \qquad\qquad □$$

Proposition 7.4 *The Cartan canonical connection of the conformal Hamiltonian metric (7.1) has two effective d-curvatures:*

$$R^l_{i1k} = \frac{\partial A^l_{i1}}{\partial x^k} - \frac{\partial \Gamma^l_{ik}}{\partial t} + A^r_{i1}\Gamma^l_{rk} - \Gamma^r_{ik}A^l_{r1}, \quad R^l_{ijk} = \mathfrak{R}^l_{ijk}.$$

Proof A Cartan canonical connection on the dual 1-jet space $J^{1*}(\mathbb{R}, \mathbb{R}^2)$ generally has five local d-tensors of curvature. For our particular Cartan canonical connection (7.9), these reduce only to two (the other three are zero):

$$R^l_{i1k} = \frac{\delta A^l_{i1}}{\delta x^k} - \frac{\delta \Gamma^{l}_{ik}}{\delta t} + A^r_{i1}\Gamma^l_{rk} - \Gamma^r_{ik}A^{l}_{r1} = \frac{\partial A^l_{i1}}{\partial x^k} - \frac{\partial \Gamma^l_{ik}}{\partial t} + A^r_{i1}\Gamma^l_{rk} - \Gamma^r_{ik}A^l_{r1},$$

$$R^l_{ijk} = \frac{\delta \Gamma^l_{ij}}{\delta x^k} - \frac{\delta \Gamma^l_{ik}}{\delta x^j} + \Gamma^r_{ij}\Gamma^l_{rk} - \Gamma^r_{ik}\Gamma^l_{rj} = \frac{\partial \Gamma^l_{ij}}{\partial x^k} - \frac{\partial \Gamma^l_{ik}}{\partial x^j} + \Gamma^r_{ij}\Gamma^l_{rk} - \Gamma^r_{ik}\Gamma^l_{rj} = \mathfrak{R}^l_{ijk}$$

$$P^{l\ (k)}_{i1(1)} = \frac{\partial A^{l}_{i1}}{\partial p^1_k} = 0, \quad P^{l\ (k)}_{ij(1)} = \frac{\partial \Gamma^l_{ij}}{\partial p^1_k} = 0, \quad S^{l(j)(k)}_{i(1)(1)} = 0. \qquad \square$$

7.4   From Hamiltonian of Time-Dependent Coupled Oscillators to Field-Like Geometrical Models

7.4.1   Gravitational-Like Geometrical Model of Momenta

The conformal Hamiltonian metric (7.1) produces on the momentum phase space $J^{1*}(\mathbb{R}, \mathbb{R}^2)$ the adapted metrical d-tensor (*momentum gravitational potential*)

$$\mathbb{G} = h_{11}dt \otimes dt + g_{ij}dx^i \otimes dx^j + h_{11}g^{ij}\delta p^1_i \otimes \delta p^1_j,$$

where g_{jk} and g^{ij} are given by (7.3). We recall that we postulate that the momentum gravitational potential $\mathbb{G}$ is governed by the *geometrical Einstein equations*

$$Ric\,(C\Gamma\,(N)) - \frac{Sc\,(C\Gamma\,(N))}{2}\mathbb{G} = \mathcal{K}\mathbb{T}, \tag{7.10}$$

where:

- $Ric\,(C\Gamma\,(N))$ is the Ricci d-tensor associated with the Cartan canonical linear connection (7.9);
- $Sc\,(C\Gamma\,(N))$ is the *scalar curvature*;
- $\mathcal{K}$ is the Einstein constant and $\mathbb{T}$ is an intrinsic momentum *stress-energy* d-tensor of matter.

Therefore, using the adapted basis of vector fields, we can locally describe the global geometrical Einstein equations (7.10). Consequently, some direct computations lead to

Proposition 7.5 *The Ricci tensor of the Cartan canonical connection of the conformal Hamiltonian metric (7.1) has the following two effective Ricci d-tensors:*

$$R_{11} := K_{11} = 0, \qquad\qquad R_{1i} = R^1_{1i1} = 0,$$
$$R_{i1} = R^r_{i1r}, \qquad\qquad R_{ij} = R^r_{ijr} = \mathfrak{R}^r_{ijr} := \mathfrak{R}_{ij},$$
$$R_{i(1)}^{(j)} := -P_{i(1)}^{(j)} = -P^{r\ (j)}_{ir(1)} = 0, \quad R_{(1)1}^{(i)} := -P_{(1)1}^{(i)} = -P^{i\ (r)}_{r1(1)} = 0,$$

$$R_{1(1)}^{(j)} := -P_{1(1)1}^{1(j)} = 0,$$
$$R_{(1)(1)}^{(i)(j)} := -S_{(1)(1)}^{(i)(j)} = -S^{i(j)(r)}_{r(1)(1)} = 0,$$
$$R_{(1)j}^{(i)} := -P_{(1)j}^{(i)} = -P^{i\ (r)}_{rj(1)} = 0.$$

Corollary 7.1 *The scalar curvature of the Cartan canonical connection of the conformal Hamiltonian metric (7.1) is given by the following formula:*

$$Sc(C\Gamma(N)) = g^{ij} R_{ij} = g^{ij} \mathfrak{R}_{ij} := \mathfrak{R}.$$

Corollary 7.2 *The geometrical momentum Einstein-like equations produced by the conformal Hamiltonian metric (7.1) are locally described by*

$$\begin{cases} -\dfrac{\mathfrak{R}}{2} h_{11} = \mathcal{K}\mathbb{T}_{11}, \\[2mm] \mathfrak{R}_{ij} - \dfrac{\mathfrak{R}}{2} g_{ij} = \mathcal{K}\mathbb{T}_{ij}, \\[2mm] -\dfrac{\mathfrak{R}}{2} h_{11} g^{ij} = \mathcal{K}\mathbb{T}^{(i)(j)}_{(1)(1)}, \\[2mm] 0 = \mathbb{T}_{1i}, \quad R_{i1} = \mathcal{K}\mathbb{T}_{i1}, \quad 0 = \mathbb{T}^{(i)}_{1(1)}, \\[2mm] 0 = \mathbb{T}^{(j)}_{i(1)}, \quad 0 = \mathbb{T}^{(i)}_{(1)1}, \quad\quad 0 = \mathbb{T}^{(i)}_{(1)j}. \end{cases}$$

The *geometrical momentum conservation-like laws* produced by the conformal Hamiltonian metric (7.1) are postulated by the following formulas:

$$\left[\frac{\mathfrak{R}}{2} \right]_{/1} = R^r_{1|r}, \quad \left[\mathfrak{R}^r_j - \frac{\mathfrak{R}}{2} \delta^r_j \right]_{|r} = 0, \quad \left[\frac{\mathfrak{R}}{2} \delta^j_r \right] |^{(r)}_{(1)} = 0,$$

where "$_{/1}$", "$_{|r}$", and "$|^{(r)}_{(1)}$" are the local covariant derivatives induced by the Cartan canonical connection $C\Gamma(N)$, and we have

$$R^i_1 = g^{iq} R_{q1}, \quad \mathfrak{R}^i_j = g^{iq} \mathfrak{R}_{qj}.$$

7.4.2   Geometrical Momentum Electromagnetic-Like 2-Form

On the momentum phase space $J^{1*}(\mathbb{R}, \mathbb{R}^2)$, the distinguished geometrical electromagnetic 2-form is defined by

$$\mathbb{F} = F^{(i)}_{(1)j} \delta p^1_i \wedge dx^j,$$

where

$$F^{(i)}_{(1)j} = \frac{h^{11}}{2} \left[g^{jk} N^{(1)}_{2\,(k)i} - g^{ik} N^{(1)}_{2\,(k)j} + \left(g^{jk} \Gamma^r_{ki} - g^{ik} \Gamma^r_{kj} \right) p^1_r \right].$$

By a direct calculation, the conformal Hamiltonian metric (7.1) produces the null momentum electromagnetic components

$$F^{(i)}_{(1)j} = 0.$$

On the Dual Jet Conformal Minkowski Hamiltonian

8

Abstract

In this chapter, we develop the distinguished Riemannian geometry (in the sense of d-torsions, d-curvatures, and momentum field-like geometrical theories) for a conformal Minkowski Hamiltonian defined in Pavlov's approach.

8.1 Introduction

In some private discussions (see also [43]), the physicist D.G. Pavlov suggested us that the well-known classical Minkowski metric can be regarded as the Finslerian metric F : $T\mathbb{R}^4 \to \mathbb{R}$, given by

$$F(y) = \sqrt{y^1 y^2 + y^1 y^3 + y^1 y^4 + y^2 y^3 + y^2 y^4 + y^3 y^4}. \tag{8.1}$$

This is because the geometrical object

$$G(y) \overset{def}{=} y^1 y^2 + y^1 y^3 + y^1 y^4 + y^2 y^3 + y^2 y^4 + y^3 y^4$$

is a quadratic form in $y = (y^1, y^2, y^3, y^4)$, whose canonical form is the Minkowski metric. In other words, using the notations

$$x = (x^1, x^2, x^3, x^4), \quad \tilde{x} = (\tilde{x}^1, \tilde{x}^2, \tilde{x}^3, \tilde{x}^4),$$

$$A = \begin{pmatrix} 1/\sqrt{6} & -1/\sqrt{3} & 1 & -1/\sqrt{6} \\ 1/\sqrt{6} & 2/\sqrt{3} & 0 & -1/\sqrt{6} \\ 1/\sqrt{6} & 0 & 0 & 3/\sqrt{6} \\ 1/\sqrt{6} & -1/\sqrt{3} & -1 & -1/\sqrt{6} \end{pmatrix},$$

© The Author(s), under exclusive license to Springer Nature Switzerland AG 2022

M. Neagu and A. Oană, *Dual Jet Geometrization for Time-Dependent Hamiltonians and Applications*, Synthesis Lectures on Mathematics & Statistics,
https://doi.org/10.1007/978-3-031-08885-8_8

and the linear coordinate transformation $^T x = A \cdot {}^T \tilde{x}$, then the induced coordinates $(\tilde{x}, \tilde{y})$ on the tangent bundle $T\mathbb{R}^4$ are $^T y = A \cdot {}^T \tilde{y}$. It follows that the Minkowski Finslerian metric (8.1) becomes

$$F(\tilde{x}, \tilde{y}) = \sqrt{(\tilde{y}^1)^2 - (\tilde{y}^2)^2 - (\tilde{y}^3)^2 - (\tilde{y}^4)^2}.$$

For such a physical reason, Balan-Neagu's monograph [44] studied the Finsler-Lagrange geometry of the *jet conformal Minkowski Lagrangian energy*

$$L : J^1(\mathbb{R}, \mathbb{R}^4) \to \mathbb{R},$$

defined by

$$L(t, x, y_1) = e^{2\sigma(x)} \cdot \left(y_1^1 y_1^2 + y_1^1 y_1^3 + y_1^1 y_1^4 + y_1^2 y_1^3 + y_1^2 y_1^4 + y_1^3 y_1^4 \right) \cdot h^{11}(t),$$

where $h_{11}(t)$ is a Riemannian metric on time manifold $\mathbb{R}$, and $\sigma(x)$ is a smooth function on spatial manifold $\mathbb{R}^4$.

Using the Legendre transformations

$$p_i^1 = \frac{\partial L}{\partial y_1^i} = e^{2\sigma(x)} \cdot \left(y_1^1 + y_1^2 + y_1^3 + y_1^4 - y_1^i \right) \cdot h^{11}(t),$$

which has the inverse

$$y_1^i = e^{-2\sigma(x)} \cdot \left(\frac{p_1^1 + p_2^1 + p_3^1 + p_4^1}{3} - p_i^1 \right) \cdot h_{11}(t),$$

then we can construct the *dual jet conformal Minkowski Hamiltonian energy*

$$H : J^{1*}(\mathbb{R}, \mathbb{R}^4) \to \mathbb{R},$$

by taking

$$H(t, x, p^1) = p_i^1 y_1^i - L = L =$$

$$= h_{11}(t) \cdot \frac{e^{-2\sigma(x)}}{3} \cdot \left[p_1^1 p_2^1 + p_1^1 p_3^1 + p_1^1 p_4^1 + p_2^1 p_3^1 + p_2^1 p_4^1 + p_3^1 p_4^1 - \right. \tag{8.2}$$

$$\left. - (p_1^1)^2 - (p_2^1)^2 - (p_3^1)^2 - (p_4^1)^2 \right].$$

8.2   The Canonical Nonlinear Connection

The spatial metrical d-tensor of the Minkowski Hamiltonian (8.2), together with its inverse, is defined by the following formulas:

$$g^{ij} = \frac{h^{11}(t)}{2} \frac{\partial^2 H}{\partial p_i^1 \partial p_j^1} = \frac{e^{-2\sigma(x)}}{6} \cdot \left(1 - 3\delta^{ij}\right) \Rightarrow g_{ij} = 2e^{2\sigma(x)} \cdot \left(1 - \delta_{ij}\right).$$

For the σ-*diagonal vector field* on $\mathbb{R}^4$, given by

$$D_\sigma = \sigma(x)\frac{\partial}{\partial x^1} + \sigma(x)\frac{\partial}{\partial x^2} + \sigma(x)\frac{\partial}{\partial x^3} + \sigma(x)\frac{\partial}{\partial x^4},$$

the corresponding divergence has the expression

$$\operatorname{div} D_\sigma = \sigma_1 + \sigma_2 + \sigma_3 + \sigma_4,$$

where $\sigma_i = \partial\sigma/\partial x^i$. Then, by direct computations, we deduce that the canonical nonlinear connection

$$N = \left(N_{1\ (i)1}^{(1)}, N_{2\ (i)j}^{(1)}\right), \tag{8.3}$$

induced by the dual jet conformal Minkowski Hamiltonian (8.2) on the dual 1-jet space $E^* = J^{1*}(\mathbb{R}, \mathbb{R}^4)$, is defined by

$$N_{1\ (i)1}^{(1)} = K_{11}^1 p_i^1 = (h^{11}/2)(dh_{11}/dt)p_i^1,$$

$$N_{2\ (i)j}^{(1)} = \frac{h^{11}}{4}\left[\frac{\partial g_{ij}}{\partial x^k}\frac{\partial H}{\partial p_k^1} - \frac{\partial g_{ij}}{\partial p_k^1}\frac{\partial H}{\partial x^k} + g_{ik}\frac{\partial^2 H}{\partial x^j \partial p_k^1} + g_{jk}\frac{\partial^2 H}{\partial x^i \partial p_k^1}\right] =$$

$$= \frac{1}{4}\left[\frac{\partial g_{ij}}{\partial x^k}\frac{\partial(h^{11}H)}{\partial p_k^1} - \frac{\partial g_{ij}}{\partial p_k^1}\frac{\partial(h^{11}H)}{\partial x^k} + g_{ik}\frac{\partial^2(h^{11}H)}{\partial x^j \partial p_k^1} + g_{jk}\frac{\partial^2(h^{11}H)}{\partial x^i \partial p_k^1}\right] =$$

$$= \frac{1}{3}\left[S^{(1)} \cdot \operatorname{div} D_\sigma - 3\langle p^{(1)}, \operatorname{grad}\sigma\rangle\right]\left(1 - \delta_{ij}\right) - \sigma_i p_j^1 - \sigma_j p_i^1,$$

where $S^{(1)} = p_1^1 + p_2^1 + p_3^1 + p_4^1$, $p^{(1)} = (p_1^1, p_2^1, p_3^1, p_4^1)$, $\operatorname{grad}\sigma = (\sigma_1, \sigma_2, \sigma_3, \sigma_4)$ and $\langle p^{(1)}, \operatorname{grad}\sigma\rangle = p_1^1\sigma_1 + p_2^1\sigma_2 + p_3^1\sigma_3 + p_4^1\sigma_4$.

8.3   Cartan Canonical Connection. d-Torsions and d-Curvatures

The canonical nonlinear connection (8.3) allows us to construct the dual *adapted bases* of d-vector and d-covector fields $\{\delta/\delta t, \delta/\delta x^i, \partial/\partial p_i^1\} \subset \mathcal{X}(E^*)$ and $\{dt, dx^i, \delta p_i^1\} \subset \mathcal{X}^*(E^*)$, where

$$\frac{\delta}{\delta t} = \frac{\partial}{\partial t} - K^1_{11} p^1_r \frac{\partial}{\partial p^1_r}, \qquad \frac{\delta}{\delta x^i} = \frac{\partial}{\partial x^i} - \overset{(1)}{N}_{(r)i} \frac{\partial}{\partial p^1_r},$$

$$\delta p^1_i = dp^1_i + K^1_{11} p^1_i dt + \overset{(1)}{N}_{(i)r} dx^r. \tag{8.4}$$

Proposition 8.1 *The Cartan canonical N-linear connection produced by the conformal Minkowski Hamiltonian metric (8.2) has the following adapted local components:*

$$C\Gamma(N) = \left(K^1_{11}, \ A^i_{j1} = 0, \ H^i_{jk} = \Gamma^i_{jk}, \ C^{i(k)}_{j(1)} = 0 \right), \tag{8.5}$$

where $\Gamma^i_{jk} = \sigma_k \delta^i_j + \sigma_j \delta^i_k - g^{ir} \sigma_r g_{jk}$.

Proof The adapted components of Cartan canonical connection are given by the following formulas:

$$A^i_{j1} = \frac{g^{il}}{2} \frac{\delta g_{lj}}{\delta t} = \frac{g^{il}}{2} \frac{d g_{lj}}{dt} = 0,$$

$$H^i_{jk} = \frac{g^{ir}}{2} \left(\frac{\delta g_{rj}}{\delta x^k} + \frac{\delta g_{rk}}{\delta x^j} - \frac{\delta g_{jk}}{\delta x^r} \right) = \frac{g^{ir}}{2} \left(\frac{\partial g_{rj}}{\partial x^k} + \frac{\partial g_{rk}}{\partial x^j} - \frac{\partial g_{jk}}{\partial x^r} \right) =$$

$$= \sigma_k \delta^i_j + \sigma_j \delta^i_k - \sigma^i g_{jk} := \Gamma^i_{jk},$$

$$C^{j(k)}_{i(1)} = -\frac{g_{ir}}{2} \left(\frac{\partial g^{jr}}{\partial p^1_k} + \frac{\partial g^{kr}}{\partial p^1_j} - \frac{\partial g^{jk}}{\partial p^1_r} \right) = 0,$$

where $\sigma^i = g^{ir} \sigma_r$. □

Proposition 8.2 *The Cartan canonical connection of the conformal Hamiltonian metric (8.2) has only one effective d-torsion, namely,*

$$R^{(1)}_{(r)ij} = - \mathcal{R}^k_{rij} p^1_k,$$

where

$$\mathcal{R}^l_{ijk} = \frac{\partial \Gamma^l_{ij}}{\partial x^k} - \frac{\partial \Gamma^l_{ik}}{\partial x^j} + \Gamma^s_{ij} \Gamma^l_{sk} - \Gamma^s_{ik} \Gamma^l_{sj}.$$

Proof The Cartan canonical connection on the dual 1-jet space $J^{1*}(\mathbb{R}, \mathbb{R}^4)$ generally has six effective local *d*-tensors of torsion. In our particular case, these reduce only to one (by computations, the other five cancel):

$$T^r_{1j} = -A^r_{j1} = 0, \quad P^{r(j)}_{i(1)} = C^{r(j)}_{i(1)} = 0,$$

$$P^{(1)\ (j)}_{(r)1(1)} = \frac{\partial N^{(1)}_{(r)1}}{\partial p^1_j} + A^j_{r1} - \delta^j_r H^1_{11} = 0, \quad P^{(1)\ (j)}_{(r)i(1)} = \frac{\partial N^{(1)}_{(r)i}}{\partial p^1_j} + \Gamma^j_{ri} = 0,$$

$$R^{(1)}_{(r)1j} = \frac{\delta N^{(1)}_{(r)1}}{\delta x^j} - \frac{\delta N^{(1)}_{(r)j}}{\delta t} = 0,$$

$$R^{(1)}_{(r)ij} = \frac{\delta N^{(1)}_{(r)i}}{\delta x^j} - \frac{\delta N^{(1)}_{(r)j}}{\delta x^i} = -\mathcal{R}^k_{rij} p^1_k. \qquad \Box$$

Proposition 8.3 *The Cartan canonical connection of the conformal Hamiltonian metric (8.2) has only one effective d-curvature:* $R^l_{ijk} = \mathcal{R}^l_{ijk}$.

Proof A Cartan canonical connection on the dual 1-jet space $J^{1*}(\mathbb{R}, \mathbb{R}^4)$ generally has five local d-tensors of curvature. In our particular case, these reduce only to one (the other four are zero). $\qquad \Box$

8.4   Momentum Field-Like Geometrical Models

8.4.1   The Gravitational-Like Geometrical Model

The Hamiltonian momentum gravitational theory produced by the Minkowski Hamiltonian (8.2) relies on the momentum gravitational h-potential $\mathbb{G}$ on the dual 1-jet space E^*, locally defined by

$$\mathbb{G} = h_{11}dt \otimes dt + g_{ij}dx^i \otimes dx^j + h_{11}g^{ij}\delta p^1_i \otimes \delta p^1_j.$$

For studying the corresponding local geometrical Einstein-like equations (together with their momentum conservation laws) in the adapted basis $\{X_A\} = \{\delta/\delta t, \delta/\delta x^i, \partial/\partial p^1_i\}$, we consider the Cartan canonical connection $C\Gamma(N) = (H^1_{11}, 0, \Gamma^i_{jk}, 0)$. Taking into account the expressions of its adapted curvature d-tensors, we get

Theorem 8.1 *The Ricci tensor* $\mathrm{Ric}(C\Gamma(N))$ *produced by the Minkowski Hamiltonian (8.2) is characterized only by one effective local adapted Ricci d-tensor:*

$$\mathcal{R}_{ij} = \mathcal{R}^r_{ijr}.$$

The scalar curvature $Sc(C\Gamma(N))$ of the Cartan connection produced by the Minkowski Hamiltonian (8.2) is given by $Sc(C\Gamma(N)) = \mathcal{R} = g^{ij}\mathcal{R}_{ij}$. Consequently, by particularizing the geometrical Einstein-like equations and the momentum conservation laws that govern an arbitrary time-dependent Hamilton space H^n, we find:

Theorem 8.2 *The local **geometrical Einstein-like equations**, that govern the momentum gravitational potential of the space produced by the Minkowski Hamiltonian (8.2), have the form*

$$
\begin{cases}
\mathcal{R}_{ij} - \dfrac{\mathcal{R}}{2}g_{ij} = \mathcal{K}\mathbb{T}_{ij}, \\
0 = \mathbb{T}_{1i}, \quad 0 = \mathbb{T}_{i1}, \quad 0 = \mathbb{T}^{(i)}_{(1)1}, \quad -\mathcal{R}h_{11} = 2\mathcal{K}\mathbb{T}_{11}, \\
0 = \mathbb{T}^{(j)}_{1(1)}, \, 0 = \mathbb{T}^{(j)}_{i(1)}, \, 0 = \mathbb{T}^{(i)}_{(1)j}, \, -\mathcal{R}h_{11}g^{ij} = 2\mathcal{K}\mathbb{T}^{(i)(j)}_{(1)(1)},
\end{cases}
\tag{8.6}
$$

where $\mathbb{T}_{AB}$, $A, B \in \left\{1, i, \overset{(i)}{\scriptstyle(1)}\right\}$ *are the adapted components of the momentum stress-energy d-tensor of matter* $\mathbb{T}$, *and* $\mathcal{K}$ *is the Einstein constant.*

As a consequence, by putting $\mathcal{R}^r_j = g^{rs}\mathcal{R}_{sj}$, then the *momentum conservation laws* of the geometrical Einstein-like equations (8.6) take the form

$$
\left[\mathcal{R}^r_j - \frac{\mathcal{R}}{2}\delta^r_j\right]_{|r} = 0.
$$

8.4.2 The Electromagnetic-Like Geometrical Model

By direct calculations, our geometrical electromagnetic-like theory produced by the Minkowski Hamiltonian (8.2) relies on the following geometrical results:

Proposition 8.4 *The **metrical deflection d-tensors** produced by the Minkowski Hamiltonian (8.2) are given by the following formulas:*

$$
\Delta^{(i)}_{(1)1} = -h_{11}g^{ik}A^r_{k1}p^1_r = 0,
$$

$$
\Delta^{(i)}_{(1)j} = h_{11}g^{ik}\left[-N^{(1)}_{(k)j} - H^r_{kj}p^1_r\right] = 0, \quad \vartheta^{(i)(j)}_{(1)(1)} = h_{11}g^{ij}.
$$

Proposition 8.5 *The distinguished momentum electromagnetic 2-form produced by the Minkowski Hamiltonian (8.2) is given by the null 2-form*

$$
\mathbb{F} = F^{(i)}_{(1)j}\delta p^1_i \wedge dx^j + f^{(i)(j)}_{(1)(1)}\delta p^1_i \wedge \delta p^1_j \equiv 0,
$$

where

$$F^{(i)}_{(1)j} = \frac{1}{2}\left[\Delta^{(i)}_{(1)j} - \Delta^{(j)}_{(1)i}\right] = 0, \quad f^{(i)(j)}_{(1)(1)} = \frac{1}{2}\left[\vartheta^{(i)(j)}_{(1)(1)} - \vartheta^{(j)(i)}_{(1)(1)}\right] = 0. \tag{8.7}$$

In conclusion, our geometrical electromagnetic-like theory produced by the Minkowski Hamiltonian (8.2) is trivial.

References

1. Abraham, R., Marsden, J.E.: Foundations of Mechanics. Benjamin, New York (1978)
2. Fomenko, A.T.: Symplectic Geometry. Methods and Applications. Gordon and Breach Publishers, New York (1988)
3. Weinstein, A.: Symplectic geometry. Bull. Amer. Math. Soc. **5**(1), 1–13 (1981)
4. Atanasiu, G.: The invariant expression of Hamilton geometry. Tensor N.S. **47**(3), 225–234 (1988)
5. Atanasiu, Gh., Klepp, F.C.: Nonlinear connections in cotangent bundle. Publ. Math. Debrecen **39**(1–2), 107–111 (1991)
6. Miron, R.: Hamilton geometry. An. Şt. "Al. I. Cuza" Univ. Iaşi **35**, 33–67 (1989)
7. Miron, R., Hrimiuc, D., Shimada, H., Sabău, S.V.: The Geometry of Hamilton and Lagrange Spaces. Kluwer Academic Publishers, Dordrecht (2001)
8. Olver, P.J.: Applications of Lie Groups to Differential Equations. Springer, New York (1986)
9. Balan, V., Neagu, M.: Jet Single-Time Lagrange Geometry and Its Applications. Wiley Inc, Hoboken (2011)
10. Atanasiu, G., Neagu, M., Oană, A.: The Geometry of Jet Multi-Time Lagrange and Hamilton Spaces. Applications in Theoretical Physics. Fair Partners, Bucharest (2013)
11. Miron, R.: Hamilton geometry. An. Şt. "Al. I. Cuza" Univ. Iaşi **35**, 33–67 (1989)
12. Miron, R., Hrimiuc, D., Shimada, H., Sabău, S.V.: The Geometry of Hamilton and Lagrange Spaces. Kluwer Academic Publishers, Dordrecht (2001)
13. Atanasiu, G.: The invariant expression of Hamilton geometry. Tensor N.S. **47**(3), 225–234 (1988)
14. Atanasiu, Gh., Klepp, F.C.: Nonlinear connections in cotangent bundle. Publ. Math. Debrecen **39**(1–2), 107–111 (1991)
15. Neagu, M., Oană, A.: Dual jet geometrical objects of momenta in the time-dependent Hamilton geometry. "Vasile Alecsandri" Univ. Bacău, Sci. Stud. Res. Ser. Math. Inform. **30**(2), 153–164 (2020)
16. Miron, R., Hrimiuc, D., Shimada, H., Sabău, S.V.: The Geometry of Hamilton and Lagrange Spaces. Kluwer Academic Publishers, Dordrecht (2001)
17. Oană, A., Neagu, M.: On dual jet N-linear connections in the time-dependent Hamilton geometry. Ann. Univ. of Craiova - Math. and Comput. Sci. Ser. **48**(1), 98–111 (2021)

© The Editor(s) (if applicable) and The Author(s), under exclusive license to Springer
Nature Switzerland AG 2022
M. Neagu and A. Oană, *Dual Jet Geometrization for Time-Dependent Hamiltonians and Applications*, Synthesis Lectures on Mathematics & Statistics,
https://doi.org/10.1007/978-3-031-08885-8

18. Neagu, M., Oană, A., Balan, V.: Dual jet h-normal N-linear connections in time-dependent Hamilton geometry. The XV-th Int. Conf. "Differ. Geom. - Dyn. Syst." (DGDS-2021), 26–29 August 2021 * ONLINE * [Bucharest, Romania]. BSG Proc. **29**, 68–73 (2022)
19. Balan, V., Neagu, M.: Ricci and deflection d-tensor identities on the dual 1-jet space $J^{1*}(\mathbb{R}, M)$. In: Proceedings of the XIII-th International Virtual Research-to-Practices Conference "Innov. Teach. Tech. in Phys. and Math., Vocat. and Mech. Train.", pp. 195–197. Mozyr State Pedagogical University named after I.P. Shamyakin, Belarus (2021)
20. Oană, A., Balan, V., Neagu, M.: Local Bianchi identities in the time-dependent Hamilton geometry on dual 1-jet spaces. In: Proc. of the XIV-th International Virtual Research-to-Practices Conference "Innov. Teach. Tech. in Phys. and Math., Vocat. and Mech. Train.", pp. 269–272. Mozyr State Pedagogical University named after I.P. Shamyakin, Belarus (2022)
21. Miron, R., Hrimiuc, D., Shimada, H., Sabău, S.V.: The Geometry of Hamilton and Lagrange Spaces. Kluwer Academic Publishers, Dordrecht (2001)
22. Neagu, M., Balan, V., Oană, A.: Dual jet time-dependent Hamilton geometry and the least squares variational method. U.P.B. Sci. Bull., Ser. A **84**, Iss. 2, 129–144 (2022)
23. Miron, R., Hrimiuc, D., Shimada, H., Sabău, S.V.: The Geometry of Hamilton and Lagrange Spaces. Kluwer Academic Publishers, Dordrecht (2001)
24. Miron, R.: Hamilton geometry. An. Şt. "Al. I. Cuza" Univ. Iaşi **35**, 33–67 (1989)
25. Atanasiu, Gh., Neagu, M., Oană, A.: The Geometry of Jet Multi-Time Lagrange and Hamilton Spaces. Applications in Theoretical Physics. Fair Partners, Bucharest (2013)
26. Neagu, M., Balan, V., Oană, A.: Dual jet time-dependent Hamilton geometry and the least squares variational method. U.P.B. Sci. Bull., Ser. A **84**, Iss. 2, 129–144 (2022)
27. Udrişte, C.: Geometric Dynamics. Kluwer Academic Publishers, Dordrecht (2000)
28. Neagu, M., Udrişte, C.: From PDE systems and metrics to multi-time field theories and geometric dynamics. West Univ. Timişoara. Semin. Mech. **79** (2001)
29. Ferrara, M., Niglia, A.: Market competition via geometric dynamics. BSG Proc. **8**, 53–59 (2003)
30. Udrişte, C., Ferrara, M., Opriş, D.: Economic Geometric Dynamics. Geometry Balkan Press, Bucharest (2004)
31. Neagu, M.: Riemann-Lagrange geometry for dynamical system concerning market competition. Bull. Transilvania Univ. Braşov. Ser. III: Math., Inform., Phys. **11**(60), (1), 99–106 (2018)
32. Balan, V., Neagu, M., Oană, A., Ovsiyuk, E.M.: A geometrization on dual 1-jet spaces of the time-dependent Hamiltonian of electrodynamics. Bull. Transilvania Univ. Braşov. Ser. III: Math. Comput. Sci. **2**(64), (1), 15–22 (2022)
33. Landau, L.D., Lifshitz, E.M.: Physique Théoretique. **1.** Mécanique. Éditions Mir, Moscou (1982) (in French)
34. Landau, L.D., Lifshitz, E.M.: Physique Théoretique. **2.** Théorie des Champ. Éditions Mir, Moscou (1989) (in French)
35. Miron, R.: Lagrange geometry. Math. Comput. Model. **20**(4/5), 25–40 (1994)
36. Neagu, M.: Riemann-Lagrange Geometry on 1-Jet Spaces. Matrix Rom, Bucharest (2005)
37. Miron, R., Anastasiei, M.: The Geometry of Lagrange Spaces: Theory and Applications. Kluwer Academic Publishers, Dordrecht (1994)
38. Miron, R.: Hamilton geometry. An. Şt. "Al. I. Cuza" Univ. Iaşi **35**, 33–67 (1989)
39. Miron, R., Hrimiuc, D., Shimada, H., Sabău, S.V.: The Geometry of Hamilton and Lagrange Spaces. Kluwer Academic Publishers, Dordrecht (2001)
40. Atanasiu, Gh., Neagu, M., Oană, A.: The Geometry of Jet Multi-Time Lagrange and Hamilton Spaces. Applications in Theoretical Physics. Fair Partners, Bucharest (2013)
41. Raeisi-Dehkordi, H., Neagu, M.: On the geometry of conformal Hamiltonian of the time-dependent coupled harmonic oscillators. Stud. Univ. Babeş-Bolyai Math. **59**(3), 385–391 (2014)

42. Macedo, D.X., Guedes, I.: Time-dependent coupled harmonic oscillators. J. Math. Phys. **53**, 052101 (2012)
43. Pavlov, D.G.: Four-dimensional time. Hypercomplex Numb. Geom. Phys. **1**(1), 31–39 (2004)
44. Balan, V., Neagu, M.: Jet Single-Time Lagrange Geometry and Its Applications. Wiley Inc, Hoboken (2011)

Index

© The Editor(s) (if applicable) and The Author(s), under exclusive license to Springer
Nature Switzerland AG 2022
M. Neagu and A. Oană, *Dual Jet Geometrization for Time-Dependent Hamiltonians
and Applications*, Synthesis Lectures on Mathematics & Statistics,
https://doi.org/10.1007/978-3-031-08885-8